T0220824

SpringerBriefs in Applied Sciences and Technology

SpringerBriefs present concise summaries of cutting-edge research and practical applications across a wide spectrum of fields. Featuring compact volumes of 50 to 125 pages, the series covers a range of content from professional to academic.

Typical publications can be:

- A timely report of state-of-the art methods
- An introduction to or a manual for the application of mathematical or computer techniques
- A bridge between new research results, as published in journal articles
- A snapshot of a hot or emerging topic
- An in-depth case study
- A presentation of core concepts that students must understand in order to make independent contributions

SpringerBriefs are characterized by fast, global electronic dissemination, standard publishing contracts, standardized manuscript preparation and formatting guidelines, and expedited production schedules.

On the one hand, **SpringerBriefs in Applied Sciences and Technology** are devoted to the publication of fundamentals and applications within the different classical engineering disciplines as well as in interdisciplinary fields that recently emerged between these areas. On the other hand, as the boundary separating fundamental research and applied technology is more and more dissolving, this series is particularly open to trans-disciplinary topics between fundamental science and engineering.

Indexed by EI-Compendex, SCOPUS and Springerlink.

More information about this series at http://www.springer.com/series/8884

Seyed Mehrshad Parvin Hosseini ·
Aydin Azizi

Big Data Approach to Firm Level Innovation in Manufacturing

Industrial Economics

 Springer

Seyed Mehrshad Parvin Hosseini
Faculty of Business
Sohar University
Sohar, Oman

Aydin Azizi
School of Engineering,
Computing and Mathematics
Oxford Brookes University
Oxford, UK

ISSN 2191-530X   ISSN 2191-5318 (electronic)
SpringerBriefs in Applied Sciences and Technology
ISBN 978-981-15-6299-0   ISBN 978-981-15-6300-3   (eBook)
https://doi.org/10.1007/978-981-15-6300-3

This Springer imprint is published by the registered company Springer Nature Singapore Pte Ltd.
The registered company address is: 152 Beach Road, #21-01/04 Gateway East, Singapore 189721,
Singapore

Contents

Chapter 1
An Introduction on Models of Innovation and Analytical Frame Work

Abstract Mobility of technological changes driven by innovation is inevitably changing the form of the world rapidly so that firms cannot catch up with the fast-changing innovative technologies. This study addresses firm level innovation issues and provides an overview of current firm level innovation in developing industries. We showcasing situation of firm level innovation among manufacturing firms since the important of level of innovation has been underestimated in previous literature. The introductory section investigated several aspects of firm level innovation: the factors that influence the decision to invest in innovation; the extent of innovation; factors characterizing an innovating firm; the types of innovation and the factors that drive and enable them. A conceptual model and an associated cost-benefit framework was developed to explain a firm's decision to invest in innovation. Then we provide details on the main drivers and enablers of innovation activities faced industrial developing countries.

Keywords Technology transfer · Firm level innovation · Manufacturing · Cost benefit analysis of innovation

1.1 Introduction

The role of innovation in development has long been of interest to economists [1–5]. Innovation activities not only result in higher production but also create technology that is transferable [6]. One of the key components of financial success in industrial firms is the extent of their innovation [7]. Innovation is often being linked to higher productivity and economic growth [8–10]. Innovations are inevitable they are a pervasive force in our organizations and in our society. Firms leverage innovation to sustain success, others do not [11].

More recently, the interest has shifted from the macro impact of innovation on the economy to factors that motivate innovation at the firm level [12–15]. Hobday [16] defines firm-level innovation as the successful introduction of a new or improved

© The Author(s), under exclusive license to Springer Nature Singapore Pte Ltd. 2020
S. M. Parvin Hosseini and A. Azizi, *Big Data Approach to Firm Level
Innovation in Manufacturing*,
SpringerBriefs in Applied Sciences and Technology,
https://doi.org/10.1007/978-981-15-6300-3_1

product, process or service to the marketplace. In order to capture incremental innovation that occurs outside formal R&D activities and from 'behind the technology frontier' defined by firms in advanced countries, he broadened his definition to include any product, process or service new to the firm, and not confined only to those new to the world or marketplace.

In recognizing the importance of innovation, Lall [17] pointed out that in the fast changing global market, successful manufacturing companies that are able to sustain their competitive advantage are the ones with faster technology adoption abilities and those involved in successful research and development that enables them to produces a flow of innovative new products over time. Firms that do not adopt these methods will fall behind competitors who do. Firms are increasingly dependent on innovation to meet greater customer requirements, to provide better service, and to face increased competitive pressures.

Innovation needs new technology and technology development takes time and requires continuous investments by firms in themselves [18–20]. However, there are firm level attributes and institutions outside the firm that can nurture the process of technology development and its conversion to innovation. Institutions outside the firm can provide important support by strengthening the science base and producing adequately trained human resources. In Asia, South Korea and Taiwan stand out as examples of countries where factors within the firms and institutions outside it has come together to foster firm-level innovation that has allowed them to catch up with firms in advanced economies [21, 22]. Evidence in the literature exist in terms of industrial developed countries tend to have higher research and development while less industrial developed countries have not enough research and development [23, 24].

There was therefore interest in determining if technology was being transferred within the manufacturing sector dominated by multinationals (MNCs) [25–27]. The modes of transfer and the linkages that have been built up between MNCs and local supporting firms [18, 28–30]. Innovation data in the developing countries are suffering from lack of consistency and changes of innovation terminologies [12]. Gault [31, 32] recommends harmonizing the definition of innovation across countries for a better policy implication. Possible reason behind unharmonized data is the nature of innovation in developing countries which is mostly incremental and less radical. Firm innovation definitions are mostly based on OSLO manual or BOGOTA manual in Southeast Asian countries. Due to difficulties in and financial constrains in developing countries there are debates on whether or not product and process innovations are happening in real sense.

This chapter is divided into three sections. The first section argues the role of eco-innovation a relatively new term that has been introduced after the social environmental concerns and sustainable development theories. Further in that section reviews the literature with respect to the definitions and main models of innovation. It shows that the existing models do not provide much insight into the factors motivating and sustaining firm-level innovation. The second section surveys the empirical literature on firm-level innovation and identifies the main correlates of innovation.

The survey reveals that in most cases the correlates have been introduced into econometric estimation models in an ad hoc way without explaining how they might fit into a coherent framework that would deepen the understanding of firm-level innovation. The final section addresses this lack by suggesting a conceptual model into which the key correlates can be fitted. This model is then used as the basis for the analytical framework underlying the empirical part of this study. In order to demonstrate the model analysis several regression-based models of the global enterprise surveys on innovation will be analysed. The next part of the review consists of set of six Southeast Asian countries profiles describing the state and dynamics of their national innovation systems.

1.2 The Role of Eco-innovation

Although debates exist in defining innovation itself the term sustainable innovation or eco-innovation has been introduced in recent years which combines the environment, social and financial aspects of innovation [33–35]. Carrillo-Hermosilla et al. [36] defines eco-innovation as innovation that improves environmental performance. European Commission [37] for instance defines the eco-innovation through responsibility of firms towards use of natural resources which also includes consumption of energy.

To paraphrase eco-innovation refers to all forms of innovation technological and non-technological that create business opportunities and benefit the environment by preventing or reducing their impact, or by optimising the use of resources. Eco-innovation is closely linked to the way we use our natural resources, to how we produce and consume and also to the concepts of eco-efficiency and eco-industries. Eco-innovation encourages a shift among manufacturing firms from "end-of-pipe" solutions to "closed-loop" approaches that minimise material and energy flows by changing products and production methods bringing a competitive advantage across many businesses and sectors.

Eco-innovation is any innovation resulting in significant progress towards the goal of sustainable development, by reducing the impacts of our production modes on the environment, enhancing nature's resilience to environmental pressures, or achieving a more efficient and responsible use of natural resources.

Research shows that eco-innovative companies of all sizes are growing, on average, at a rate of 15% a year, even at the time of economic rescissions. In fact, the survival rate of eco-innovative firms during economic rescission are much higher [38]. Small and medium-sized enterprises (SMEs) are particularly becoming responsive to eco-innovation due to their adaptability and flexibility, and as contributors of as much as 70% of GDP and two-thirds of formal employment in developing and emerging economies, eco-innovation is potentially a key driver of a resource efficient economy.

Despite the fact that literature on the determinants of green innovation or eco-innovation is rare just like innovation the term eco-innovation has gained attention in industrial sections mostly in agri-food and chemical industries [39]. For measuring eco-innovation there are no single method or approach is likely to be sufficient. Even any patent covering environment and sustainability cannot directly measure eco-innovations but can act as intermediate indicator for inventions. Terms like eco-innovation, green innovation and environmental innovation has been used interchangeably in the literature. Industries that search for new opportunities and involve themselves in continues collaboration are more prone to develop eco-innovations. Since the approach of innovation and eco-innovation and determinants that shape the two concepts are almost identical, we mainly concentrate of the factors that leads to higher firm level innovation. This is based on the hypothesis that if a firm has the innovation capabilities, they also have higher tendency to have eco-innovation.

1.3   Indicators and Measurement of Innovation

Existing models of innovation do not provide much insight into the factors motivating and sustaining firm-level innovation. The work of Joseph Schumpeter [40] has greatly influenced modern ideas of innovation. He viewed innovation as a dynamic process in which new technologies replace the old, a process he labeled "creative destruction". He divided innovation into two types, incremental and radical. For Schumpeter, radical innovations created major disruptive changes, whereas incremental innovations continuously advanced the process of change.

Since then various definitions of innovation have appeared in the literature throughout the years (see, for example, [41–45] but over time the definitions given in the *Oslo Manual* [46] has become a standard point of reference. The Manual classified innovation into four types: product, process innovation, marketing and organization.

i. **Product innovation** refers to new knowledge or technology used to introduce new or improved products or services. An improved product (or service) is an existing product (or service) whose performance has been significantly enhanced or upgraded.
ii. **Process innovation** is the implementation of new or significantly improved production or delivery method to generally increase productivity.
iii. **Marketing innovation** covers all new marketing methods, product designs, packaging, product placements, product promotions and/or or pricing.
iv. **Organization innovation** refers to the implementation of new organizational methods in the firm's business practices, work place organization or external relations.

It was emphasized that the innovation should be new to the firm though not necessarily new to the market. The very recent definition of innovation has been changed and according to Oslo Manual [24]: "An innovation is a new or improved product or process, or combination thereof, that differs significantly from the unit's previous products or processes and that has been made available to potential users (product) or brought into use by the unit (process)." It was also immaterial whether the innovation was developed by the main firm or by another enterprise. However, changes of a solely aesthetic nature and the mere selling of innovations produced and developed entirely by other firms were not counted as innovation.

Shumpeter's concepts of incremental and radical innovation have also been further refined in the broader literature [47–50]. The former refers to small improvements in the areas of product, process, marketing or organization that helps a firm maintain its competitive position over time. The latter, on the other hand, refers to any innovation that completely replaces existing products, processes or marketing and organizational structures. Further the concepts of implementation of innovation and diffusion has been added to the definition of innovation [50].

Further studies [48, 51, 52] accumulate firm's sales revenues from products that are new to the domestic and international markets to measure radical innovation, and uses a firm's sale revenue from products that are new to the firm itself to measure incremental innovation. While distinguishing incremental from radical innovation may be useful for pedagogical purposes the distinctions are often blurred in practice [53, 54].

For the scope of this study, innovation was defined as new or significant changes to goods or services; production or delivery methods. While marketing and organizational innovations are important, the data used in the study preclude their inclusion. Innovation is further divided into three types, following a classification used by the World Bank [55–57] introduced by Yevgeny Kuznetsov and consistent with the data collected by the collaborative efforts of the World Bank and the local Department of Statistics in various countries. Innovating firms were classified as adopters, adapters or creators based on the innovative activity they were primarily involved in.

1.4 Models of Innovation

There is no single general theory or model of innovation. In a review of models of firm-level innovation Hobday [16] identified five generations of such 'models', based on the work of Rothwell [58]. Godin reviewed the history of innovation [59] and gave more sight to the these theories. They basically purport to describe how and from where firm-level innovation is motivated. While these models did not develop sequentially, they all appear to be a description of the ideas regarding innovation that was prevalent in different periods of time. The five generations of models are:

1. Technology Push (1950s to mid-1960s)
2. Market Pull (mid-1960s–1970s)
3. Coupling models (mid 1970s–1980s)
4. Integrated model (early 1980s–1990)
5. Systems integration and networking model (post-1990).

However, the main threads of these models are actually captured in two models of innovation. The Technology Push and Market Pull models are just variations of the old Linear Model of innovation while the remaining models are variations of the Chain-Linked Model. The discussion will therefore focus on these two. The ideas behind each of these models are discussed in turn.

1.4.1 The Linear Model of Innovation

The models that Hobday referred to as 'Technology Push' and 'Market Pull' are essentially variations of the old Linear Model of Innovation. The source of this Linear Model, according to Godin [59, 60] is "nebulous, having never been documented" but it is "one of the first (conceptual) frameworks developed for understanding the relation of science and technology to the economy".

In articulating the relationship between science and innovation, it begins by separating science into basic and applied sciences. It assumes research is motivated by basic science and not by economic objectives. It is born from a curiosity and the urge to know or understand some phenomenon. Research is therefore undertaken by academics or public scientists located outside industry [61]. However, the basic research is often generating massive data, related theories, and potential explanations for technological phenomenon. Most of basic research outputs can be useful in applied research [62–64]. In this model, basic research starts the push for innovation.

In the second step, applied research draws on discoveries from basic research to develop specific applications that may be useful for industries or other fields. This can be done by procuring directly the knowledge or through collaboration with other research institutions. Firms are therefore seen as important agents who take discoveries from basic research and channel them to applications. Once an application or invention is obtained, then the huge cost of development is recovered through the final step of commercialization (Fig. 1.1).

Since the progress from basic research to commercialization was posited to move in a linear fashion, the model became known as Linear Model.

The Linear Model of innovation has been heavily criticized [60, 65–67] but the major limitation in the context of the present study is that sheds no light on factors that might nurture or sustain research at the level of the firm.

Despite its shortcomings, the model may have been a fairly accurate description of an era when firms had limited resources to undertake research of any kind, let alone basic research. Almost all research was publicly funded and was undertaken outside industry; firm-level applied research and innovation was largely motivated by the

Fig. 1.1 Linear model of innovation (technology push) [60]

Fig. 1.2 Linear model of innovation (market pull) [60]

'push' from research done exogenously. Even today, the linear model may be relevant in understanding innovation in at least some industries like aerospace.

In the 'Market Pull" variation, the linear sequential process remained, except that the catalyst for innovation was no longer seen as originating from sources completely unrelated to the needs of the firm as posited in the original 'push' version. Rather, the demand for firm-level innovation emerges from the markets that firms serve and research responds to the market needs of customers. The demand in the marketplace therefore pulled research in directions necessary to serve these needs (Fig. 1.2).

The "market pull" version of the linear model at least offered one insight on what might motivate firm innovation in a firm. It suggests that the changing needs and demands of the customers pressure firms to innovate.

1.4.2 The Chain Linked Model

The alternative view of linear innovation model is called Chain Linked Model which also explains the relationship between science and technological innovation. It is reportedly "the most cited non-linear innovation model" and was developed by Stephen Kline of Stanford University [67]. Others tried to improve the Chain Linked Model [68].

The Chain Linked Model no longer views innovation as only occurring in industry or outside it. Neither was it posited to move in a linear fashion from research to commercialization. Rather it admits that research occurs in both places concurrently with much of the basic research occurring outside industry and much applied research occurring within it [66, 69]. Schot [69] also highlight the new model as firms are more dealing with environmental challenges such as the sustainable development goals.

In the choice of the areas of research, the model admits the possibility that scientists may be influenced not only by curiosity but also by what technology can do. It therefore concedes that scientists can also be influenced by the needs of industries.

In this framework the relationship between scientists, science and technologies become complex extremely complex with feedback loops from later to earlier stages and vice versa (Fig. 1.3). The central idea of chain-linked model is that scientific research may actually occur in separate institutions such as universities or public laboratories and that it generates substantial amounts of scientific knowledge. However, innovation in firms occurs in tandem but without needing basic science as their

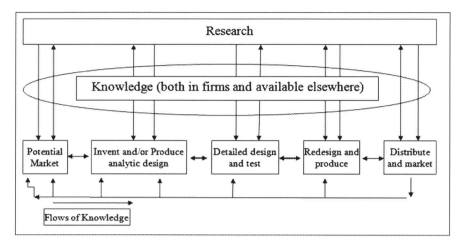

Fig. 1.3 Chain linked model [65]

starting point. Rather, it is driven by potential demand for new products in the market and they utilize all available knowledge and expertise that already exists in the company to satisfy the potential market. This approach is less costly and firms do not start from scratch every time. From this beginning, in analytical design firms can move to detailed designs, testing, redesigning etc before final distribution in the market. Ideally, a firm tries to move from the lower, left part of Fig. 1.3 to the right side of the figure without ever looking into the science. However, if any of these steps in the lower half of Fig. 1.3 fails to move the firm from the left side to right side, it would prompt firms to look back to existing scientific knowledge in order to gain a theoretical understanding of the problem and hopefully find a solution as well.

1.5 Conclusion

We explained variation of innovation models mainly linear model of innovation and Chain Linked model. The Chain Linked Model retains the view that the potential demand for new products motivates innovation in firms but notes that they will depend largely on applied research to do so because basic research is too expensive to start up. The third, fourth and fifth generation models described by Hobday [16] may be viewed as further elaborations of the basic Chain Link Model. There are other similar models of innovation, which have been described by Schot [69] and others like Boons et al. [34] and Mahjoubi [70]. It is clear from the discussion that these models are not very useful in understanding or explaining firm level innovation. While the later models suggest that firm level innovation is a response to market needs, they ignore other possible drivers of innovation. Furthermore, these models do not discuss factors that support and enable innovation in a firm.

References

1. T. Hagerstrand, *Innovation Diffusion as a Spatial Process* (University Chicago Press, Chicago, USA, 1968), p. 334
2. Z.J. Acs, D.B. Audretsch, Innovation in large and small firms. Am. Econ. Rev. **23**(1), 678–690 (1990)
3. G. Dosi, Sources, procedures, and microeconomic effects of innovation. J. Econ. Lit. **26**(3), 1120–1171 (1988)
4. L.K. Mytelka, Rethinking development: a role for innovation networking in the 'other two-thirds'. Futures **25**(6), 694–712 (1993)
5. R.R. Nelson, *National Innovation Systems: A Comparative Study*, vol. 2, no. 8 (Oxford University Press, 1993), p 45
6. R. Griffith et al., Innovation and productivity across Four European countries. Oxford Rev. Econ. Policy **22**(4), 483–498 (2006)
7. B. Bigliardi, The effect of innovation on financial performance: A research study involving SMEs. Innovation **15**(2), 245–255 (2013)
8. B.H. Hall, F. Lotti, J. Mairesse, Innovation and productivity in SMEs: empirical evidence for Italy. Small Bus. Econ. **33**(1), 13–33 (2009)
9. P. Mohnen, B.H. Hall, Innovation and productivity: an update. Eurasian Bus. Rev. **3**(1), 47–65 (2013)
10. B.H. Hall, *Innovation and Productivity* (National Bureau of Economic Research, 2011)
11. J.E. Ettlie, E.M. Reza, Organizational integration and process innovation. Acad. Manag. J. **35**(4), 795–827 (1992)
12. X. Cirera, S. Muzi, Measuring innovation using firm-level surveys: evidence from developing countries. Res. Policy **49**(3), 103912 (2020)
13. X. Sun, H. Li, V. Ghosal, Firm-level human capital and innovation: evidence from China. China Econ. Rev. **59**, 101388 (2020)
14. E. Kremp, J. Mairesse, *Knowledge Management, Innovation, and Productivity: A Firm Level Exploration Based on French Manufacturing CIS3 Data* (National Bureau of Economic Research, 2004), p. 10237
15. R. Evangelista, A. Vezzani, The economic impact of technological and organizational innovations. A firm-level analysis. Res. Policy **39**(10), 1253–1263 (2010)
16. M. Hobday, Firm-level innovation models: perspectives on research in developed and developing countries. Technol. Anal. Strateg. Manag. **17**(2), 121–146 (2005)
17. S. Lall, Competitiveness indices and developing countries: an economic evaluation of the global competitiveness report. World Dev. **29**(9), 1501–1525 (2001)
18. D. Keppler, Characterisation of innovations within the multi-level perspective with diffusion typology of innovations: a fruitful combination. J. Innov. Manag. **7**(2), 15–37 (2019)
19. J. De Cubas, Technology transfer and the developing nations. Counc. Am. **5**(7), 23 (1974)
20. R. Chester Goduscheit, R. Faullant, Paths toward radical service innovation in manufacturing companies—a service-dominant logic perspective. J. Prod. Innov. Manag. **35**(5), 701–719 (2018)
21. K.J. Fields, *Enterprise and the State in Korea and Taiwan* (Cornell University Press, 2019)
22. C.-H. Huang, C.-H. Yang, Persistence of innovation in Taiwan's manufacturing firms. Taiwan Econ. Rev. **38**(2), 199–231 (2010)
23. OECD, *Innovation in Southeast Asia* (2013)
24. OECD, *OECD Science, Technology and Innovation Outlook 2018* (2018)
25. D. Audretsch, R. Caiazza, Technology transfer and entrepreneurship: cross-national analysis. J. Technol. Transf. **41**(6), 1247–1259 (2016)
26. J.R. Holm, et al., Labor mobility from R&D-intensive multinational companies: implications for knowledge and technology transfer. J. Technol. Transf. (2020)
27. S. Narayanan, Y.W. Lai, Technological maturity and development without research: the challenge for Malaysian manufacturing. Dev. Change **31**(2), 435–457 (2000)

28. N. Hovhannisyan, Technology gap and international knowledge transfer: new evidence from the operations of multinational corporations. Eastern Econ. J. **45**(4), 612–638 (2019)
29. G. Batra, H. Tan, *SME technical efficiency and its correlates: cross-national evidence and policy implications* (World Bank, Washington, DC, 2003), p. 32
30. A. Giroud, *Transnational Corporations, Technology, and Economic Development: Backward Linkages and Knowledge Transfer in South East Asia*, vol. 1 (Edward Elgar Publishing, United Kingdom, 2003), p. 358
31. F. Gault, Social impacts of the development of science, technology and innovation indicators. Technology and Innovation Indicators (January 25, 2011) (2011)
32. F. Gault, Developing a science of innovation policy internationally, in *Science of Science Policy: A Handbook* (Stanford University Press, Stanford, 2011)
33. P. Ekins, Eco-innovation for environmental sustainability: concepts, progress and policies. IEEP **7**(2), 267–290 (2010)
34. F. Boons et al., Sustainable innovation, business models and economic performance: an overview. J. Clean. Prod. **45**, 1–8 (2013)
35. C. Montalvo, F. Diaz-Lopez, F. Brandes, *Eco-innovation Opportunities in Nine Sectors of the European Economy*. European Sector Innovation Watch. European Commission, Directorate General Enterprise and Industry, Brussels (2011)
36. J. Carrillo-Hermosilla, P.R. del González, T. Könnölä, *What is Eco-Innovation?* in *Eco-innovation* (Springer, 2009), pp. 6–27
37. European Commission, *Eco-Innovation the Key to Europe's Future Competitiveness* (2020)
38. S.E. Pérez, A.S. Llopis, J.A.S. Llopis, The determinants of survival of Spanish manufacturing firms. Rev. Ind. Organ. **25**(3), 251–273 (2004)
39. E.M. García-Granero, L. Piedra-Muñoz, E. Galdeano-Gómez, Measuring eco-innovation dimensions: the role of environmental corporate culture and commercial orientation. Res. Policy. **49**(8),104028 (2020)
40. J.A. Schumpeter, *The Theory of Economic Development: An Inquiry into Profits, Capital, Credit, Interest, and the Business Cycle*. 10th edn., vol. 55 (Harvard College, United State of America, 1934), p. 255
41. P.F. Drucker, *Innovation and Entrepreneurship: Practice and Principles*, 2th edn., R. Edition, vol. 3 (Routledge, Oxford, UK, 1985), p. 253
42. D. Leonard-Barton, W.C. Swap, *When Sparks Fly: Igniting Creativity in Groups* (Harvard Business Press, United State of America, 1999), p. 242
43. H. Li, K. Atuahene-Gima, Product innovation strategy and the performance of new technology ventures in China. Acad. Manag. J. **44**(6), 1123–1134 (2001)
44. F. Zhao, Exploring the synergy between entrepreneurship and innovation. Int. J. Entrepreneurial Behav. Res. **11**(1), 25–41 (2005)
45. S. Massa, S. Testa, Innovation and SMEs: misaligned perspectives and goals among entrepreneurs, academics, and policy makers. Technovation **28**(7), 393–407 (2008)
46. OECD, *Oslo Manual, Guidelines for Collecting and Interpreting Innovation Data*, in *A joint publication of OECD and Eurostat, Statistical Office of the European Communities* (Organization for Economic Co-operation and Development, Luxembourg, 2005), p. 202
47. D.A. Norman, R. Verganti, Incremental and radical innovation: design research vs. technology and meaning change. Des. Issues **30**(1), 78–96 (2014)
48. P. Ritala, P. Hurmelinna-Laukkanen, Incremental and radical innovation in coopetition—the role of absorptive capacity and appropriability. J. Prod. Innov. Manag. **30**(1), 154–169 (2013)
49. J.E. Ettlie, W.P. Bridges, R.D. O'Keefe, Organization strategy and structural differences for radical versus incremental innovation. Manage. Sci. **30**(6), 682–695 (1984)
50. OECD, *Oslo Manual 2018 Guidelines for Collecting, Reporting and Using Data on Innovation*. Letöltve, p. 15. https://read.oecd-ilibrary.org/science-and-technology/oslo-manual-2018_9789 264304604-en (2018)
51. B. Forés, C. Camisón, Does incremental and radical innovation performance depend on different types of knowledge accumulation capabilities and organizational size? J. Bus. Res. **69**(2), 831–848 (2016)

52. B.S. Tether, A. Tajar, The organisational-cooperation mode of innovation and its prominence amongst European service firms. Res. Policy **37**(4), 720–739 (2008)
53. V. Chiesa et al., Exploring management control in radical innovation projects. Eur. J. Innov. Manag. **12**(4), 416–443 (2009)
54. R.B. Bouncken et al., Coopetition in new product development alliances: advantages and tensions for incremental and radical innovation. Br. J. Manag. **29**(3), 391–410 (2018)
55. T.W. Bank, *Innovation Policy: A Guide for Developing Countries* (World Bank Publications, 2010)
56. World Bank, Malaysia firm competitiveness, investment climate, and growth, Report No. 26841-MA, in *Based on the Productivity and Investment Climate Survey Conducted Between December 2002 and May 2003 with Reference Period of 1999–2001* (Poverty Reduction Economic Management and Financial Sector Unit: East Asia and Pacific Region), p. 212
57. World Bank, Malaysia productivity and investment climate assessment update, Report No. 49137-MY, in *Based on the Productivity and Investment Climate Survey Conducted Between 2006 and 2007 with Reference Period of 2006*, ed. by J.W. Adams, et al. (World Bank Poverty Reduction and Economic Management Sector Unit, East Asia and Pacific Region, 2009), p. 260
58. R. Rothwell, Successful industrial innovation: critical factors for the 1990s. R&D Manag. **22**(3), 221–240 (1992)
59. B. Godin, The Elgar companion to innovation and knowledge creation, in *A Conceptual History of Innovation* (Edward Elgar Publishing, 2017)
60. B. Godin, The linear model of innovation: the historical construction of an analytical framework. Sci. Technol. Human Values **31**(6), 639–667 (2006)
61. D. Foray, F. Lissoni, Chapter 6—University research and public–private interaction, in *Handbook of the Economics of Innovation*, ed. by H.H. Bronwyn, R. Nathan (North-Holland, 2010), pp. 275–314
62. M. Coccia, Competition between basic and applied research in the organizational behaviour of public research labs. J. Econ. Lib. **5**(2), 118–133 (2018)
63. F. Lissoni, Academic patenting in Europe: a reassessment of evidence and research practices. Ind. Innov. **20**(5), 379–384 (2013)
64. O. Sorenson, L. Fleming, Science and the diffusion of knowledge. Res. Policy **33**(10), 1615–1634 (2004)
65. S.J. Kline, Innovation is not a linear process. Res. Manag. **28**(4), 36–45 (1985)
66. S.J. Kline, N. Rosenberg, An overview of innovation, in *The Positive Sum Strategy: Harnessing Technology for Economic Growth*, ed. by R. Landau, N. Rosenberg (National Academy Press, United State of America, 1986), pp. 275–305
67. S. Kline, C.L. Model, *Non-linear Models of Technological Innovation the Mapping of Innovation* (University of Texas, 1997), p. 26
68. J.-P. Micaëlli et al., How to improve Kline and Rosenberg's chain-linked model of innovation: building blocks and diagram-based languages. J. Innov. Econ. Manag. **3**, 59–77 (2014)
69. J. Schot, W.E. Steinmueller, Three frames for innovation policy: R&D, systems of innovation and transformative change. Res. Policy **47**(9), 1554–1567 (2018)
70. D. Mahjoubi, *Non-linear Models of Technological Innovation: The Mapping of Innovation*, vol. 2, no. 3 (University of Texas, 1997), p. 26

Chapter 2
The Correlates of Firm-Level Innovation

Abstract The lack of a satisfactory theory or framework to understand firm-level innovation has caused large body of literature examining the effect of different variables on firm-level innovation in an ad hoc fashion. The primary usefulness of this approach is that it has helped identify potential correlates of innovation. After surveying the literature, the correlates of firm-level innovation have been divided into two main groups for this study: factors that motivate or drive innovation and factors that support or enable innovation. The latter has been further subdivided into three groups of factors. One, firm-level characteristics that facilitate innovation; two, factors that lower the cost of innovation; and three, public policies that nurture innovation. Each of these groups is discussed in turn.

Keywords Knowledge cumulativeness · Technology transfer · Facilitator of innovation · Productivity · Market competition · Literature review

2.1 Introduction

Two well-known models of innovation (Linear Model and Chain-Linked Model) do not provide much insight into factors motivating and sustaining firm level innovation. In fact, shortage of studies exists in term of the drivers of innovation. Competition is a key driver of innovation. Marx [1], in his book on capitalistic production, recognized that competition forces firms to replace old modes of technology with new ones; this can be rephrased to mean that competition drives innovation. Even a firm that has no ambitions beyond serving the market it currently operates in must innovate to stay in business, if there are competitors. Nevertheless, more proactive firms that seek to increase revenues by venturing into new markets or by offering new services or even attempting to respond to new market needs cannot avoid innovation [2–4]. However, Xia [5] and Oliveira and Fujiwara [4] noted that while some degree of competition is necessary to spur faster innovation, there could be cases where too much competition makes innovation unprofitable.

© The Author(s), under exclusive license to Springer Nature Singapore Pte Ltd. 2020   13
S. M. Parvin Hosseini and A. Azizi, *Big Data Approach to Firm Level Innovation in Manufacturing*,
SpringerBriefs in Applied Sciences and Technology,
https://doi.org/10.1007/978-981-15-6300-3_2

Several proxies have been used in the literature to capture the effect of competition on innovation. These include export orientation, market concentration and differences in manufacturing subsectors.

2.2   Conditions in Individual Subsectors

Competitive pressures vary, not only according to the market's firms operate in but also according to the structure of the subsectors, they operate in. Subsectors that are more competitive are believed to spur greater innovation activities [5–8], while firms operating in industries with short product cycles may be expected to be more innovative. The empirical results on these notions are however inconclusive.

In Malaysian context for instance, it has been suggested that the electronics subsector in Penang is more competitive because of shorter product cycles relative to the electrical products subsector [9, 10]. Similarly, R&D was found also to be greater in firms involved in high-tech activities in Israel [11, 12] and Japan [13] although Lee and Lee [14] found that Malaysian SMEs in sectors with sophisticated technology were less likely to innovate.

In Britain, Pavitt [6] and later other researchers [15] observed that the more innovative firms were in the sophisticated electronic and chemical sectors. In Malaysia, technology transfer, and by implication innovation, had reached higher stages in the competitive electronics and electrical (EE) subsector in Penang, relative to other subsectors [9, 10]. Another studies arrived at the same conclusion noting that innovation was most evident in the EE subsector [16, 17]. Other studies also suggest that the intensity of market competition or monopolization in some sectors may affect the outcome of innovation strategies but without consensus on the direction of association [5, 18, 19].

2.3   Export Orientation

Export-orientation often functions as a proxy for competitive pressures; a firm operating only in the foreign market or in both the foreign and domestic markets is likely to face greater competitive pressures than firms that operate only in the domestic market. Thus, a positive correlation between export orientation and innovation may be expected though it is not always clear in which direction the causation runs. Lall [20, 21] for example, believed that technological capability building was the driver of export competitiveness and growth. Wagner [22], on the other hand, argued that firms participating in international markets are exposed to more intense competition and are forced to improve (innovate) faster than firms that sell their products purely in domestic markets. Further Fassio [23] confirmed positive effect of exporting on innovation in France, Germany, Italy, Spain, and UK.

Most empirical studies report a positive association between operating in export markets and innovation. Another category of firm level studies link international-ization, innovativeness and growth and found they were strongly correlated and possibly mutually reinforcing in technology-intensive small firms [24–26]. Rasiah and Krishnan [27] found a significant relationship between export orientation and products and processes innovations while in Germany a similar finding was reported with respect to exporting firms in services relative to non-exporting firms in the sector [28, 29]. In Latin America, manufacturing firms operating in export markets were reportedly more innovative than firms operating in the domestic market [30, 31].

Somewhat throughout the time some of these studies reported contradictory find-ings for instance by Lee and Lee [14] with respect to 2002 Malaysian data. Drawing on a sample of 239 firms from the National Survey of Innovation-3 (NSI-3) for the reference period 2000–2001, they ran simple Probit (and Logit) regressions to iden-tify significant characteristics of firms that innovate from firms that did not. SMEs were divided in small (less than 50 workers), medium-sized (50–249 workers) and large firms (250 workers or more). Restricting the discussion to the first two groups, they found that export-orientation was associated with a lower probability of inno-vation among medium-sized firms and not significant among the small firms. No reasons were afforded for these findings.

Smaller sub-sectorial studies of varying quality also exist. Ng and Thiruchelvam [32] surveyed innovation among smaller wooden furniture manufacturers in Malaysia. Their findings, based on responses from 97 firms (out of 300 that were canvassed) showed that 72% of the respondents were doing some kind of innovation. The majority of firms were small or medium-sized with innovation being negligible among micro firms. Most innovating firms were export-dependent and younger in age.

2.4 Market Concentration

If competition spurs innovation, a subsector characterized by market segmentation and a few dominant firms is less likely to nurture innovative firms. However, Schum-peter [33] advanced the idea that firms in highly concentrated markets will be more innovative; it follows from this that policies that limit competition might adversely affect innovative activities.

The empirical evidence again remains inconclusive; a review of the empirical liter-ature up to the late 1970s by Kamien and Schwartz in 1982 established this (cited in [34]). Later studies continue into give contradictory results. Heredia [35] and Miguel-Benavente [36] for example, found that larger firms and firms with higher market shares in their industry had higher R&D intensities in Chile. Similarly, Hosseini and Narayanan [37] found that Malaysian SMEs in highly concentrated markets had a higher probability of innovating. In contrast, studies such as Blundell et al. [38] and Geroski [39] found evidence against the Schumpeterian view. Gayle [34] suggests that the lack of support for Schumpeter's view is because in most studies' innovation was measured using simple patent counts; he used citation-weighted patents counts and showed a significant and positive association between industrial concentration and innovation.

2.5   Firm Characteristics that Facilitate Innovation

The firm-level characteristics frequently referred to in the literature are firm size, age of firm, equity ownership and organizational characteristics.

2.5.1   Firm-Size

Firm size has traditionally been viewed as a positive influence on innovation [33, 40]. Bigger firms have the resources to invest in innovation and enjoy scale economies that justify the large investments in innovation infrastructure [31, 35, 41]. However, as firms grow large, their R&D becomes less efficient. Levin and Reiss [42] reviewed the empirical evidence and observed that though economies of scale and scope may exist, they are often exhausted long before a firm becomes very large.

Analysis of Enterprise Survey data which was conducted by World Bank for numerous countries has produced contradictory results in terms of innovation. For instance, in the case of Malaysian manufacturing firms Lee [43] and Hosseini and Narayanan [37] noted that large Malaysian firms are more likely to innovate compared to small firms. On the other hand, Rasiah [44] studied a small sample of 151 small and medium sized enterprises (SMEs) and their access to finance in Malaysia. An incidental finding relevant here was the inverse relationship between firm size and participation in R&D.

The R&D statistics collected through the World Bank's industrial survey [45] also showed that SMEs are credit constrained and cannot fulfil normal collateral requirements. Large firms on the other hand have performed better in terms of innovation activities (as measured by dollar expenditures) than small firms. However, large firms were more export-oriented than the smaller ones so it is unclear if innovation was associated with size or export orientation.

Studies elsewhere too have suggested that firm size is not necessarily positively correlated with innovation [46–49]. These studies show that small firms are active in innovation. In Finland, Simonen and McCann [50, 51] too found that small firms were generally more innovative than larger firms were.

Smaller firms were also found to be very innovative in the service sector. Innovation is often more informal, diffused and less input intensive allowing small firms to quickly attain the critical mass necessary to sustain innovation. Additionally, the service sector offers more opportunities to small-and-medium-sized enterprises (SMEs) and is, in fact, dominated by such firms relative to the manufacturing sector [52].

2.5.2   Age of Firm

It is reasonable to expect that better-established firms are more likely to innovate since innovation is a risky venture that requires substantial capital investments, knowledge of the market preferences and the resources to translate research findings into products, processes or marketing innovation. Jefferson et al. [53], for example, argue that age proxies a firm's experience and is therefore positively related to innovation. Conversely Huergo [54] expounded that new firms entering to market tend to present the highest probability of innovation. Similarly among Spanish firms Coad et al. [55] elucidate that younger firms undertake riskier innovation activities which might have greater return or loss. Therefore, younger firms face larger performance benefits from R&D.

Debates exist among the findings of these studies for instance in a study of manufacturing firms in the UK, Criscuolo et al. [56] also found that more established firms led in innovation. Shefer and Frenkel [11] noted that younger Israeli firms were more R&D intensive than older ones.

2.5.3   Foreign Ownership

In looking at firm ownership foreign equity participation is often considered as giving the firm an edge in innovation [57]. Among Spanish firms however Diaz [58] found no significant differences between the innovation of foreign-owned firms and locally owned firms.

Among developing countries multinational corporations have spearheaded the industrialization process in many countries like Malaysia, Singapore, Indonesia and Thailand and have also served as channels for the transfer of technology, though the pace of transfer has remained a point of contention [59–61].

In Malaysian manufacturing, it is widely held that foreign–owned firms are more involved in innovation than their local counterparts [9, 61, 62]. This was true of some Latin American countries as well [31]. A large study of firms in China found that firms with foreign capital participation (and those with good access to domestic bank loans) innovate more than other firms do [63]. Another comprehensive study of the Indian firms by the National Knowledge Commission (NKC) also reported that firms with majority foreign ownership were more innovative than those with majority Indian ownership [64].

However, foreign ownership may well be capturing the effects of other conditions that spur innovation like access to technology and operating in a competitive environment. Once these effects are controlled for, foreign ownership may not be independently and significantly related to innovation [65].

2.5.4 Organizational Characteristics

Organizational factors like decentralised hierarchical structures, good communication channels among staff, attitude towards risk of management etc. have also been discussed in the literature as determining whether or not a firm innovates [66].

2.6 Factors Lowering Innovation Costs

Several factors may be expected to directly or indirectly lower the costs of innovation. Among the factors discussed in the literature are the following.

2.6.1 Access to Technology

The second set of correlates of innovation discussed in the literature refers primarily to modes of access to technology, including paying royalties to use patented technologies and outsourcing innovation. Access to technology may be expected to be positively associated with innovation because they provide a base for innovation and may lower the costs of innovation.

Access to technology can be gained by simply buying it from outside sources. Usually, the payments of royalties by firms suggest that technology has been bought. Technology can be also be gained from parent plants in the case of foreign subsidiaries located in the local economy [67–69] or through links with MNCs, in the case of domestic firms in supplier relationships with them [70–72]. Beladi et al. [73] and Bell et al. [74], for example, reported that the subcontracting links that domestic firms have with demanding MNCs improved the product quality and technological capabilities of the former. Other similar study is in line with same hypothesis [75].

Collaborative research, in particular, has been highlighted in empirical studies. Several studies suggest that firms collaborate to acquire knowledge they do not have, to gain complementary resources or finance, or to reduce innovation costs [18, 76, 77]. The study by Fu et al. [18] found that firms also seek partners with complementary expertise to gain quick access to new or leading knowledge or to learn through networking. Mairesse and Mohnen [78] observed that members of a collaborative group might benefit from intra-group knowledge spillovers and internal access to finance. Their study of the firms in European Community found a positive association between collaboration in R&D and the propensity to innovate. Some scholars have noted that inter-organizational collaboration may be an effective strategy to learn and manage intellectual capital [79].

Innovation is also facilitated through collaboration with universities or research institutions outside [80, 81] or through the receipt of technical support from outside agencies [82, 83]. Enkel et al. [84] argued that firms have to co-innovate with external sources such as universities, research and technology institutions in order to share R&D costs and reduce the risks of failure. They further asserted that once the notion of inter-organizational innovation collaboration has entered an industry, any firm, which does not participate, would experience serious competitive disadvantages.

Firms may also generate knowledge endogenously by creating and maintaining research facilities and staff. The availability of researchers is often viewed as an important factor in sustaining the growth of the firm and in keeping it abreast with competitors [18, 85]. Since R&D is a significant indicator of innovation, it is argued that firms with dedicated R&D staff may be expected to show a higher propensity to innovate [7, 86]. However, the presence of R&D staff cannot predict the *type* of innovation a firm is likely to be engaged in.

The outsourcing R&D, on the other hand, confer firms with some of the benefits of in-house innovation [87–89] without the need to undertake independent in-house innovative activities. Outsourcing R&D is a form of 'open innovation' that allows the firms to leverage on the technological capabilities of expertise outside the firm [87]. Outsourcing innovation can therefore result in firms reporting gains from innovation without engaging in innovation of their own. Un [90] stated that there is an inverse U-shaped relationship with product innovation exist when firms are learning from R&D outsourcing.

2.6.2 Skilled Workers

The lack of skilled and professional workers is often highlighted as an obstacle to innovation in the literature. Skilled workers are the main pillar of innovation-based firms. Skilled workers can motivate research and development that are transferable and can be utilized to create and initiate innovative activities. Fassio et al. [91] found skilled educated migrants in French and the UK firms will increase knowledge sharing and patenting per capita and overall positive effects on innovation. The implication is that the availability of skilled personnel will facilitate innovation [92–94]. A similar argument is framed in terms of excessive reliance on unskilled workers in the Malaysian context. Malaysian firms with a large share of unskilled immigrant workers are assumed to not engage in innovation since they are presumed to be in low-tech, labour intensive activities [62]. Among Southeast Asian countries Singapore and Malaysia are among the highest recipience of foreign skilled workers.

2.6.3 Industrial Clustering

The advantage of location in aiding innovation has been recognized in the literature [95–98]. However, the idea has largely been discussed in terms of countries. Countries with well-developed national innovation systems are attractive to innovative firms. However, Porter and Stern [99] draw attention to locational advantages offered by specific locations within a country that may be favourable to innovation. These locations are often areas where industrial clusters have developed over time. They highlight four cluster-specific environments for innovation: the presence of high-quality and specialised inputs; a context that encourages investment together with intense local rivalry; pressure from sophisticated local demand; and the presence of related and supporting industries. They argue that affirm within a cluster can more easily source new components, services, machinery and other elements necessary to innovate. The complementary relationships involved in innovating are also more easily forged among participants who are nearby. Reinforcing these advantages for innovation is the competitive pressure, peer pressure, customer pressure and constant comparison that is inherent within a cluster [100]. Empirical evidence in support is however inconclusive. Moderate support comes from Taiwan's semiconductor industry [101]. However, other studies do not find much evidence of cluster-specific advantages promoting innovation [102].

For instance, despite the lack of convincing evidence, Malaysia's Second Industrial Master Plan (1996–2005), embraced the idea of a cluster development strategy. How consistently it was implemented is, however, unclear. Some studies have suggested that SMEs located in Penang Malaysia have leveraged on their links with multinationals and have gone much further in the industrial upgrading process than in other locations [59].

2.6.4 Government Policies and Incentives

Since innovation considered as a public good research and development is always suffering from under provision or efficiency of funding innovation either by government or private institution. Innovative projects also tend to suffer from under provision and inefficiency more specifically in developing countries. Current world level of innovation growth is insufficient for generating welfare specifically in helping developing or underdeveloped nations. The impact of government funding on research projects has been examined by several scholars. For instance, Zhou et al. [103] studied how project-based government funding can help the innovation cultural among creative industries in China. They divided the project-based government funding into two major categories of central-government-funded projects (CGFPs) and local-government-funded research projects (LGFPs). Controlling for incremental and radical innovation they found in both categories CGFP and LGFP have an inverted U-shaped effect on firms' incremental innovation but government funding could not impact in the same manner for radical innovation.

Incentives or public support for private R&D is justified on at least three grounds [104]. First, is the 'market-failure' argument, which suggests that because the private returns to R&D are lower than the social returns, without public support R&D will always remain at a socially sub-optimal level. Second, firms may simply lack the resources to finance important R&D, or even if well endowed, may be unable to do it fast enough or adequately. Third, firms without internal resources may find it expensive to finance R&D by borrowing from external sources because the rate of return required by an investor utilizing his own funds is often lower than the rate of return expected by the external lender.

Incentives for innovation are becoming popular worldwide; a 2011 Report prepared by Price Water House Coopers (PwC) [105] on tax incentives to promote innovation in Malaysia argued that they are effective because they lower the marginal cost of innovation. It noted that 21 countries of the Organisation for Economic Co-operation and Development (OECD) have implemented fiscal incentives to hasten the pace of innovation. Even non-OECD countries such as Brazil, China, India, Singapore and South Africa are offering competitive tax incentives to raise investments in innovation [106].

A wide-ranging review of studies on the effectiveness of fiscal incentives on R&D concluded that tax incentives are useful in stimulating private R&D and raising the level of expenditure on business R&D [107]. However, in a study of 304 Malaysian manufacturing firms Rasiah and Krishnan [27] found that government incentives for innovation was positive and highly significant only in the case of larger firms (employing more than 150 workers). In Spain, a country with very generous incentives for R&D, they were again found government incentives to be significant only for larger firms while they were utilized "only randomly" by SMEs [108].

2.7 Conclusion

This study has extensively reviewed the existing literature on the parameters to determine innovation activities among nations in manufacturing firms to provide a comprehensive view. We divided these factors into two major categories of enablers and drivers of innovation. These categories were further divided to small subcategories. Three main subcategories of the drivers were conditions in individual subsector that a firm is active in, export orientation and market concentration of the manufacturing firm. Further, enablers were divided to 3 main subcategories of factors that facilitate innovation, factors that lowering the innovation cost and policy environment that a firm is active in. Each three categories were involved with some variables that also were discussed.

References

1. K. Marx, Capital: a critical analysis of capitalist production, in *Das Kapital: A Critique of Political Economy*, 2nd edn., vol. 2 (S. Sonnenschein, Lowrey, & Company, Germany), p. 356
2. G. Berkhout, D. Hartmann, P. Trott, Connecting technological capabilities with market needs using a cyclic innovation model. R&D Manag. **40**(5), 474–490 (2010)
3. B.A. Lukas, O.C. Ferrell, The effect of market orientation on product innovation. J. Acad. Mark. Sci. **28**(2), 239–247 (2000)

4. Gd Oliveira, T. Fujiwara, Intellectual property and competition as complementary policies: a test using an ordered probit model. Funpacao Getuliovergas **1**(30), 163 (2007)
5. T. Xia, X. Liu, Foreign competition, domestic competition and innovation in Chinese private high-tech new ventures. J. Int. Bus. Stud. **48**(6), 716–739 (2017)
6. K. Pavitt, Sectoral patterns of technical change: towards a taxonomy and a theory. Res. Policy **13**(6), 343–373 (1984)
7. J. Baldwin, P. Hanel, D. Sabourin, Determinants of innovative activity in Canadian manufacturing firms: the role of intellectual property rights. Analytical Stud. Branch **1**(11F0019), 38 (2000)
8. A. Trigo, Innovation patterns under the magnifying glass: firm-level latent class analysis of innovation activities in services. The University of Manchester **2**(578), 23 (2009)
9. UNDP, *Technology Transfer to Malaysia: A Study of the Electronics and Electrical Industry Goods Sector and the Supporting Firms Industries in Penang* (United Nations Development Programme, Kuala Lumpur, 1994)
10. R. Rasiah, *Institutions and Public-Private Partnerships: Learning and Innovation in Electronics Firms in Penang, Johor and Batam-Karawang* (Institutions and Economies, 2017), pp. 206–233
11. D. Shefer, A. Frenkel, R&D, firm size and innovation: an empirical analysis. Technovation **25**(1), 25–32 (2005)
12. M.A. Cusumano, A. Gawer, D.B. Yoffie, *The Business of Platforms: Strategy in the Age of Digital Competition, Innovation, and Power* (HarperCollins, New York, NY, 2019)
13. N. Wang, G. Mogi, Deregulation, Market Competition, and Innovation of Utilities: Evidence from Japanese Electric Sector. Energy Policy **111**, 403–413 (2017)
14. C. Lee, C.G. Lee, SME innovation in the Malaysian manufacturing sector. Econ. Bull. **12**(30), 1–12 (2007)
15. C. Velu, A. Jacob, Business model innovation and owner–managers: the moderating role of competition. R&D Manag. **46**(3), 451–463 (2016)
16. P.-C. Athukorala, Global productions sharing and local entrepreneurship in developing countries: evidence from Penang Export Hub, Malaysia. Asia Pac. Policy Stud. **4**(2), 180–194 (2017)
17. WB/UNDP, *Made in Malaysia: Technology Development for Vision 2020*. United Nations Development Programme/World Bank Integrated: Report Submitted to Government of Malaysia, September (1995)
18. X. Fu, H. Xiong, J. Li, Open innovation as a response to constraints and risks: evidence from China, in *Asian Economic Panel Meeting in March 19–20th, 2013* (MIT Press, Universiti Sains Malaysia (USM), Penang, Malaysia, 2013), pp. 1–23
19. A. Fosfuri, The licensing dilemma: understanding the determinants of the rate of technology licensing. Strateg. Manag. J. **27**(12), 1141–1158 (2006)
20. S. Lall, Competitiveness Indices and developing countries: an economic evaluation of the global competitiveness report. World Dev. **29**(9), 1501–1525 (2001)
21. S. Lall, Technological capabilities and industrialization. World Dev. **20**(2), 165–186 (1992)
22. M. Wagner, On the relationship between environmental management, environmental innovation and patenting: evidence from German manufacturing firms. Res. Policy **36**(10), 1587–1602 (2007)
23. C. Fassio, Export-led innovation: the role of export destinations. Ind. Corp. Change **27**(1), 149–171 (2017)
24. H.-J. Cho, V. Pucik, Relationship between innovativeness, quality, growth, profitability, and market value. Strateg. Manag. J. **26**(6), 555–575 (2005)
25. D. Keeble et al., Internationalisation processes, networking and local embeddedness in technology-intensive small firms. Small Bus. Econ. **11**(4), 327–342 (1998)
26. J.O. Rypestøl, J. Aarstad, Entrepreneurial innovativeness and growth ambitions in thick vs. thin regional innovation systems. Entrepreneurship Reg. Dev. **30**(5-6), 639–661 (2018)
27. R. Rasiah, G. Krishnan, Economic performance and technological intensities of manufacturing firms in Malaysia: does size matter? Asian J. Technol. Innov. **16**(1), 63–82 (2008)

28. B. Peters et al., Internationalisation, innovation and productivity in services: evidence from Germany, Ireland and the United Kingdom. Rev. World Econ. **154**(3), 585–615 (2018)
29. G. Licht et al., Innovation in the service sector: selected facts and some policy conclusions. Center for European Economic Research, Mannheim (ZEW) **4**(12), 134–160 (1999)
30. M. Iizuka, M. Gebreeyesus, 'Discovery' of non-traditional agricultural exports in Latin America: diverging pathways through learning and innovation. Innov. Dev. **8**(1), 59–78 (2018)
31. G. Crespi, P. Zuniga, Innovation and productivity: evidence from six Latin American countries. World Dev. **40**(2), 273–290 (2012)
32. B.-K. Ng, K. Thiruchelvam, The dynamics of innovation in Malaysia's wooden furniture industry: innovation actors and linkages. Forest Policy Econ. **14**(1), 107–118 (2012)
33. J.A. Schumpeter, *Capitalism, Socialism and Democracy*, 3rd edn., vol. 3 (Routledge, New York, United State of America, 1942, 1975 Reprint), p. 249
34. P.G. Gayle, Market concentration and innovation: new empirical evidence on the Schumpeterian hypothesis, in *University of Colorado at Boulder: Unpublished Paper* (Kansas State University, Manhattan, Kansas, 2001), p. 35
35. J.A. Heredia Pérez et al., New approach to the innovation process in emerging economies: the manufacturing sector case in Chile and Peru. Technovation **79**, 35–55 (2019)
36. J. Miguel Benavente, The role of research and innovation in promoting productivity in chile. Econ. Innov. New Technol. **15**(4–5), 301–315 (2006)
37. S.M.P. Hosseini, S. Narayanan, Adoption, adaptive innovation, and creative innovation Among SMEs in Malaysian manufacturing. Asian Econ. Pap. **13**(2), 32–58 (2014)
38. R. Blundell, R. Griffith, J.V. Reenen, Dynamic count data models of technological innovation. Econ. J. **105**(429), 333–344 (1995)
39. P.A. Geroski, Innovation, technological opportunity, and market structure. Oxford Economic Papers New Series **42**(3), 586–602 (1990)
40. J.K. Galbraith, *American Capitalism: The Concept of Countervailing Power*, 2nd edn., ed. 8, vol. 61 (Harmondsworth, Penguin, United State of America, 1970[1956]), p. 222
41. W.M. Cohen, D.A. Levinthal, Innovation and learning: the two faces of R&D. Econ. J. **99**(397), 569–596 (1989)
42. R.C. Levin, P.C. Reiss, Cost-reducing and demand-creating R&D with spillovers. National Bureau of Economic Research Cambridge. RAND J. Econ. **19**(4), 538–556 (1989)
43. C. Lee, The determinants of innovation in the Malaysian manufacturing sector: an econometric analysis at the firm level. Asean Econ. Bull. **21**(3), 319–329 (2004)
44. R. Rasiah, Financing small and medium manufacturing firms in Malaysia, in *Small and Medium Enterprises (SMES) Access to Finance in Selected East Asian Economies*, ed. by C. Harvie, S. Oum, D. Narjoko (ERIA, Jakarta, Indonesia, 2011), pp. 231–260
45. X. Cirera, W.F. Maloney, *The Innovation Paradox: Developing-Country Capabilities and the Unrealized Promise of Technological Catch-Up* (The World Bank, 2017)
46. A.M. Knott, C. Vieregger, Reconciling the firm size and innovation puzzle. Organ. Sci. **31**(2), 477–488 (2020)
47. D.B. Audretsch, et al., *Firm Size and Innovation in the Service Sector* (2018)
48. X. Fang, N.R. Paez, B. Zeng, The nonlinear effects of firm size on innovation: an empirical investigation. Econ. Innov. New Technol., 1–18 (2019)
49. W.-L. Lin et al., Does firm size matter? Evidence on the impact of the green innovation strategy on corporate financial performance in the automotive sector. J. Clean. Prod. **229**, 974–988 (2019)
50. J. Simonen, P. McCann, Innovation, R&D cooperation and labor recruitment: evidence from Finland. Small Bus. Econ. **31**(2), 181–194 (2008)
51. J. Simonen, P. McCann, Firm innovation: the influence of R&D cooperation and the geography of human capital inputs. J. Urb. Econ. **64**(1), 146–154 (2008)
52. L. Rubalcaba, The challenges for service innovation and service innovation policies. Promoting Innovation in the Services Sector. Rev. Experiences Policies **1**(6), 3–29 (2011)
53. G.H. Jefferson et al., R&D performance in Chinese industry. Econ. Innov. New Technol. **15**(4–5), 345–366 (2006)

54. E. Huergo, J. Jaumandreu, How does probability of Innovation change with firm age? Small Bus. Econ. **22**(3), 193–207 (2004)
55. A. Coad, A. Segarra, M. Teruel, Innovation and firm growth: does firm age play a role? Res. Policy **45**(2), 387–400 (2016)
56. P. Criscuolo, N. Nicolaou, A. Salter, The elixir (or burden) of youth? Exploring differences in innovation between start-ups and established firms. Res. Policy **41**(2), 319–333 (2012)
57. M. Guadalupe, O. Kuzmina, C. Thomas, Innovation and Foreign Ownership. Am. Econ. Rev. **102**(7), 3594–3627 (2012)
58. N.L. Díaz-Díaz, I. Aguiar-Díaz, P. De Saá-Pérez, Impact of foreign ownership on innovation. Eur. Manag. Rev. **5**(4), 253–263 (2008)
59. S. Narayanan, Y.W. Lai, Technological maturity and development without research: the challenge for Malaysian manufacturing. Dev. Change **31**(2), 435–457 (2000)
60. S.B. Choi, S.H. Lee, C. Williams, Ownership and firm innovation in a transition economy: evidence from China. Res. Policy **40**(3), 441–452 (2011)
61. S.M. Parvin Hosseini, Innovative capabilities among SMEs in Malaysian manufacturing: an analysis using firm-level data. N. Z. Econ. Pap. **48**(3), 257–268 (2014)
62. NEM, *The New Economic Mode for Malaysia, Part 1*, ed. by F.G.A. Centre (National Economic Advisory Council, Kuala Lumpur, Malaysia, 2010)
63. S. Girma, Y. Gong, H. Görg, Foreign direct investment, access to finance, and innovation activity in Chinese enterprises. World Bank Econ. Rev. **22**(2), 367–382 (2008)
64. NKC National Knowledge Commission, *Innovation in India, New Delhi*, ed. by K.C. Department (National Knowledge Commission Government of India, India, 2007), p. 218
65. B.M. Sadowski, G. Sadowski-Rasters, On the innovativeness of foreign affiliates: evidence from companies in The Netherlands. Res. Policy **35**(3), 447–462 (2006)
66. D. Wan, C.H. Ong, F. Lee, Determinants of firm innovation in Singapore. Technovation **25**(3), 261–268 (2005)
67. E. Mollick, A. Robb, Democratizing innovation and capital access: the role of crowdfunding. Calif. Manag. Rev. **58**(2), 72–87 (2016)
68. J.H. Dunning, Multinational enterprises and the globalization of innovatory capacity. Res. Policy **23**(1), 67–88 (1994)
69. B. Kogut, International management and strategy, in *Handbook of Strategy and Management*, ed. by A.M. Pettigrew, H. Thomas, R. Whittington (Sage, California, 2002), pp. 261–278
70. D. Ernst, L. Kim, Global production networks, knowledge diffusion, and local capability formation. Res. Policy **31**(8–9), 1417–1429 (2002)
71. R. Qiu, J. Cantwell, General purpose technologies and local knowledge accumulation—a study on MNC subunits and local innovation centers. Int. Bus. Rev. **27**(4), 826–837 (2018)
72. N. Kawai, R. Strange, A. Zucchella, Stakeholder pressures, EMS implementation, and green innovation in MNC overseas subsidiaries. Int. Bus. Rev. **27**(5), 933–946 (2018)
73. H. Beladi, A. Mukherjee, Union bargaining power, subcontracting and innovation. J. Econ. Behav. Organ. **137**, 90–104 (2017)
74. M. Bell, et al., *Aiming for 2020: A Demand Driven Perspective on Industrial Technology Policy in Malaysia*. Final Report for the World Bank and the Ministry of Science, Technology and the Environment, Malaysia. SPRU Mimeo (1995)
75. V.R. Isaac et al., From local to global innovation: The role of subsidiaries' external relational embeddedness in an emerging market. Int. Bus. Rev. **28**(4), 638–646 (2019)
76. W.H. Hoffmann, R. Schlosser, Success factors of strategic alliances in small and medium-sized enterprises—an empirical survey. Long Range Plan. **34**(3), 357–381 (2001)
77. O. Gassmann, Opening up the innovation process: towards an agenda. R&D Manag. **36**(3), 223–228 (2006)
78. J. Mairesse, P. Mohnen, Accounting for innovation and measuring innovativeness: an illustrative framework and an application. Am. Econ. Rev. **92**(2), 226–230 (2002)
79. Walter W. Powell et al., Network dynamics and field evolution: the growth of interorganizational collaboration in the life sciences. Am. J. Sociol. **110**(4), 1132–1205 (2005)

80. M.J. Nieto, L. Santamaría, The importance of diverse collaborative networks for the novelty of product innovation. Technovation **27**(6), 367–377 (2007)
81. R.D. Fitjar, A. Rodríguez-Pose, Firm collaboration and modes of innovation in Norway. Res. Policy **42**(1), 128–138 (2013)
82. B.H. Hall, F. Lotti, J. Mairesse, Innovation and productivity in SMEs: empirical evidence for Italy. Small Bus. Econ. **33**(1), 13–33 (2009)
83. B.H. Hall, *Innovation and Productivity* (National Bureau of Economic Research, 2011)
84. E. Enkel, O. Gassmann, H. Chesbrough, Open R&D and open innovation: exploring the phenomenon. R&D Manag. **39**(4), 311–316 (2009)
85. V. Van de Vrand, et al., *Open Innovation in SMEs: Trends, Motives and Management Challenges*. Technovation **29**(6), 423–437 (2009)
86. B.H. Hall, S.A. Merrill, *Research and Development Data Needs: Proceedings of a Workshop* (National Academies Press, Washington, D.C., 2005), p. 52
87. H.W. Chesbrough, *Open Innovation: The new Imperative for Creating and Profiting from Technology*, ed. 1578518377, vol. 42 (Harvard Business Press, Boston, 2003), p. 231
88. O. Jones, Innovation management as a post-modern phenomenon: the outsourcing of pharmaceutical R&D. Br. J. Manag. **11**(4), 341–356 (2000)
89. J. Hsuan, V. Mahnke, Outsourcing R&D: a review, model, and research agenda. R&D Manag. **41**(1), 1–7 (2011)
90. C.A. Un, A. Rodríguez, Learning from R&D outsourcing vs. learning by R&D outsourcing. Technovation **72-73**, 24–33 (2018)
91. C. Fassio, F. Montobbio, A. Venturini, Skilled migration and innovation in European industries. Res. Policy **48**(3), 706–718 (2019)
92. R. Tiwari, S. Buse, Barriers to innovation in SMEs: can the internationalization of R&D mitigate their effects? in *Proceedings of the First European Conference on Knowledge for Growth: Role and Dynamics of Corporate R&D-CONCORD* (Hamburg University of Technology, Seville, Spain, 2007)
93. D. Crown, A. Faggian, J. Corcoran, Foreign-born graduates and innovation: evidence from an Australian skilled visa program ☆, ☆☆, ★, ★★. . Res. Policy 103945 (2020)
94. P. Aghion, et al., *Innovation, firms and wage inequality*. CEPR Working Paper (2017)
95. C.G. Culem, The locational determinants of direct investments among industrialized countries. Eur. Econ. Rev. **32**(4), 885–904 (1988)
96. S. Hong, S. Kim, Locational determinants of Korean manufacturing investments in the European Union. Hitotsubashi J. Econ. **44**(1), 91–103 (2003)
97. E.G. Carayannis et al., Location and innovation capacity in multilevel approaches: editorial note. J. Knowl. Econ. **7**(4), 837–841 (2016)
98. A. Alventosa et al., Location and innovation optimism: a behavioral-experimental approach. J. Knowl. Econ. **7**(4), 890–904 (2016)
99. M. Porter, S. Stern, Innovation: location matters. MIT Sloan Manag. Rev. Summer **42**(4), 28–36 (2001)
100. N.J. Lowe, L. Wolf-Powers, Who works in a working region? Inclusive innovation in the new manufacturing economy. Reg. Stud. **52**(6), 828–839 (2018)
101. P.J. Sher, P.Y. Yang, The effects of innovative capabilities and R&D clustering on firm performance: the evidence of Taiwan's semiconductor industry. Technovation **25**(1), 33–43 (2005)
102. F. Huber, Do clusters really matter for innovation practices in information technology? Questioning the significance of technological knowledge spillovers. J. Econ. Geogr. **12**(1), 107–126 (2012)
103. Zhou, J., et al., The more funding the better? The moderating role of knowledge stock on the effects of different government-funded research projects on firm innovation in Chinese cultural and creative industries. Technovation (2018)
104. D. Czarnitzki, A. Fier, Do innovation subsidies crowd out private investment? Evidence from the German service sector. Publikationen von Forscherinnen und Forschern des ZEW **2**(4), 27 (2002)

105. PWC, Price Water House Coopers Innovation, in © *2010 PricewaterhouseCoopers LLP* (Price Water House Coopers, PWC Main Office, USA, 2011), pp. 1–68
106. P. Larédo, C. Köhler, C. Rammer, The Impact of fiscal incentives for R&D, in *Handbook of Innovation Policy Impact* (Edward Elgar Publishing, 2016)
107. C. Köhler, P. Laredo, C. Rammer, The impact and effectiveness of fiscal incentives for R&D. Manchester Institute of Innovation Research **1**(12), 37 (2012)
108. B. Corchuelo, E. Martínez-Ros, The effects of fiscal incentives for R&D in Spain. Open Access Publications from Universidad Carlos III de Madrid **09–23**(02), 32 (2009)

Chapter 3
Firm-Level Innovation: A Conceptual Model to Firm Level Innovation

Abstract The review of literature confirms the presence of a large body of empirical work on the correlates of firm-level innovation but there is no conceptual framework that ties these correlates together into a coherent whole. The conceptual model provides the basis for developing an analytical framework to understand the role of the drivers and enablers in encouraging innovation. The underlying idea is that the firm does a cost-benefit calculation to make two decisions: (i) whether or not to invest in innovation; (ii) and, if it decides to do so, the level of innovation to be achieved. Based on the cost and benefit analysis of firm level innovation a conceptual model was therefore developed to better understand the links of these correlates to firm level innovation

Keywords Cost benefit analysis · HH index · Competition · Types of innovation · Literature review

3.1 Introduction

Adam Smith's in his book "Wealth of Nations" mentioned innovation is the product of division of labour, which leads to economic growth. However, Smith's idea about innovation was an amateur by our modern standards, his view of innovation as a professional activity was indeed ahead of its time. Closely related to the concept of technology is not only division of labour but also the notion of invention and innovation. One of the distinctions that associate with invention is "newness" and "usefulness" refers to putting an invention in to the practice that associate with innovation [1]. Innovation then is the first application of invention a technological change in production and the beginning of a diffusion process [2]. The Innovation measurement studies in industrial sectors includes measuring knowledge production function introduce by Solow [3] who explained there are other factor rather than labour and capital can affect the output returns. Solow [3] made his famous residuals, called technology latent dependent variable in Cobb Douglas production function. Other

© The Author(s), under exclusive license to Springer Nature Singapore Pte Ltd. 2020   27
S. M. Parvin Hosseini and A. Azizi, *Big Data Approach to Firm Level
Innovation in Manufacturing*,
SpringerBriefs in Applied Sciences and Technology,
https://doi.org/10.1007/978-981-15-6300-3_3

researcher also studied the influence of more productive labour on faster technological progress [4]. After Solow, some of the evolutionary economics like Freeman [5] emphasized to the role of institutions and institutional change in advancing innovation. Perhaps the most convincing direct evidence in favour of viewing industrial innovation as the engine of growth comes from the work of economic historians.

Critical role of firm level innovation and its impact on economic growth is uncontested. Innovative firms constantly trying to growth in terms of efficiency and productivity. Hall et al. [6], Hall [7] highlighted the nexus of productivity and innovation. Research and development were considered as the main factor of innovation that can boost the productivity Hall et al. [6–8]. Other studies emphasized on the relationship of size of establishment measured by number of full-time employees and innovation [9]. Innovation can be viewed in several dimensions in terms of geographic boundaries or sector of technology. The review of literature confirms the presence of a large body of empirical work on the correlates of firm-level innovation but there is no conceptual framework that ties these correlates together into a coherent whole. Several factors have been identified to impact innovation yet less is known about the main factors influencing innovation as a standardised formulated model. The importance of knowing these factors as a standardize framework were undermined. Evidently substantial efforts were made by countries to implement innovation surveys however, innovation data are still suffering from lack of sampling methods inconsistencies throughout years. Innovation firm level survey data in hand are still scattered and suffering from several challenges like lack of standardized methodology among countries or any existing international instruction manual [10]. Therefore, designing a new conceptual framework to predict enablers and drivers of innovation is well-placed.

General idea behind designing framework for innovation is competition. Manufacturing firms have to eventually innovate in order to protect their market share and survive in the competitive market, failing which, it will be forced out of the market. Therefore, market competition plays a vital role in forcing firms to innovate. The degree of competitive pressure will vary with different circumstances and technological environments. For instance, it is common sense to expect that a firm operating in both the domestic and foreign markets through export will face greater competition than one serving only the domestic market. A conceptual model was therefore developed to better understand the links of these correlates to firm level innovation (see Fig. 3.1). A preliminary version of this conceptual model was developed and used in Hosseini and Narayanan [11].

3.2 Driver of Innovation

The model assumes that the perceived benefits from innovation encourage it but the underlying pressure or force is the need to stay competitive. The pressure can take several forms; at the lowest level, it may be just a response to changing market demands or simply the need to maintain current market shares and revenues from

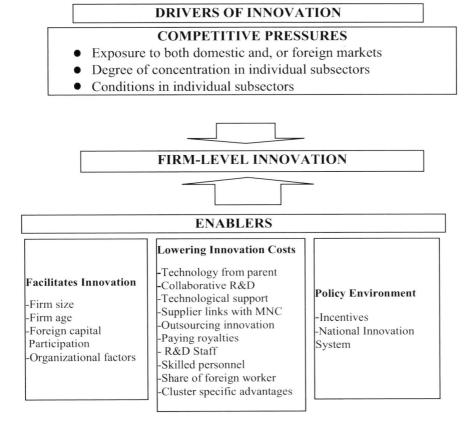

Fig. 3.1 Conceptual model of innovation—drivers and enablers of firm level innovation [11]

erosion by competitors. At another level, a proactive firm may want to expand revenues by expanding market shares at home or abroad or both. It is worth noticing that a country open regime for trade and investment in the manufacturing sector ensures a constant competitive threat for domestic firms [12]. A firm must therefore adopt new technologies (when internal innovative activity does not exist) or innovate in order to protect its market share and continue operating, or to expand its share of the market to enhance revenues. In short, in order to reap the benefits from innovation a firm must enhance its competitiveness and this is only possible through innovation [13].

The degree of competitive pressures will vary with different circumstances. For example, it is reasonable to expect that a firm operating in both the domestic and foreign markets will face greater competition than one serving only the domestic market [14]. The degree of competition is also likely to vary by subsector, with those facing shorter product cycles or in highly concentrated markets being exposed to more intense pressures to innovate [15, 16].

These factors are captured by the box at the top of the figure and identified as *drivers* of innovation. These drivers exert pressure on the firm to innovate as indicated by the arrow pointing downward to the firm.

3.3 Enablers of Innovation

The extent to which a firm is able to respond to the pressures to innovative is influenced by *enablers* of innovation. These are divided into three groups: enablers likely to facilitate engaging in innovation, enablers that may lower the cost of engaging in innovation and, finally, enablers in the overall policy environment likely to encourage innovation. The last group includes specific incentives designed by policy makers to encourage firm-level innovation by lowering its cost and, more generally, the 'national innovation systems' that encompasses a wider network of players and institutions that support innovation [17, 18]. The enablers are shown in the Fig. 3.1 as three pillars supporting firm-level innovation.

An advantage of the conceptual model is that it helps identify factors favouring innovation that are outside the control of the firm (drivers) and those within the control of the firm (the first two sets) and factors that are the outcome of joint-efforts of the firm and the government (the third set of enablers). This helps to identify policies that can be initiated by the firm and those that need the intervention of the authorities.

The conceptual model may be summarised as follows: competition drives innovation and innovation is necessary to move ahead of the competition in order to gain the perceived benefits. In this effort, the firm may be aided by enablers (or hindered by their absence). One set of enablers identified from the literature survey (firm size, age, equity composition and organizational features) may facilitate undertaking innovation. Another set discussed in the literature (various modes of access to technology) may encourage innovation by lowering the cost of engaging in it (as will be shown below). The third set mentioned in some studies (government policies and the policy environment) may help to nurture innovation by lowering the costs. It also may correct for the positive externalities associated with innovation. The possible impact of these enablers on innovation is discussed in greater detail below.

3.4 An Analytical Framework: The Cost and Benefits of Firm-Level Innovation

The conceptual model provides the basis for developing an analytical framework to understand the role of the drivers and enablers in encouraging innovation. The underlying idea is that the firm does a cost-benefit calculation to make two decisions: (i) whether or not to invest in innovation; (ii) and, if it decides to do so, the level of innovation to be achieved.

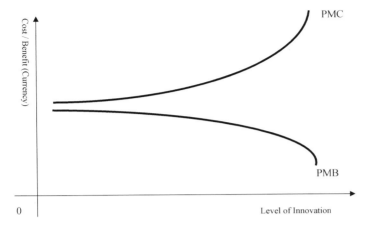

Fig. 3.2 Private marginal costs and benefits of innovation [19]

Assuming the technology underlying innovation activity in a firm is fixed in a given time period, the firm can expect to receive additional benefits from increasing the level of innovation although this will also incur further costs. The private marginal benefits (PMB) and private marginal costs (PMC) from additional levels of innovation are represented by the negatively sloped and positively sloped curves in Fig. 3.2, respectively. The horizontal axis measures the additional levels of innovation undertaken and the vertical axis measures the currency value of the associated benefits and costs.

The negatively sloped PMB curve reflects the usual assumption of diminishing returns to additional levels of innovation while the positively sloped PMC shows the rising costs of increasing levels of innovation.

3.5 The Decision to Innovate

The framework provides some insights on the question of *whether or not to invest in innovation*. In order to decide the firm will compare the PMC and PMB schedules. If the PMC is expected to remain above PMB at all levels of investment in innovation (as shown in Fig. 3.2), the firm will not innovate in-house. In this case, it is faced with three choices: (i) take the risk of not innovating at all; (ii) gain the benefit from innovation by outsourcing it; or (iii) acquire newer technology through the payment of royalties. Which of these it will opt for cannot be predicted in advance from the framework but the firm will decide based on a comparison of the net benefits to be gained from each of these choices? This aspect of whether or not to innovate is explored in further sections.

3.6 Facilitators of Innovation

The first set of enablers identified in the conceptual model (firm size, age, equity composition and organizational features) either is acquired over time or can be developed. They play no direct role in the decision to innovate, although they may have a supportive role once the decision to innovate is made. For example, once the decision to invest is made, a larger firm with greater resources or one with greater market experience may be able to sustain higher levels of innovation than a smaller firm, or a younger firm. These are therefore not discussed further in the context of the cost-benefit framework.

3.7 Costs of Innovation

Once the decision to invest in innovation is made, the firm has to determine the level of innovation to invest in. For this, the firm balances the costs involved against the benefit (revenue) gained from the additional level of investment being contemplated. If the positive externalities associated with innovation are ignored (as firms most likely will), they will choose the level of investment where the PMB = PMC (point E_1 in Fig. 3.3). This will *not* reflect the socially optimal level of innovation and in fact will be below it (or at a suboptimal level).

Note that point E_1 represents the private equilibrium level of funds invested in innovation *with the given technology in place*. (The adoption of new and more advanced technology can be expected to affect both the PMB and PMC curves.)

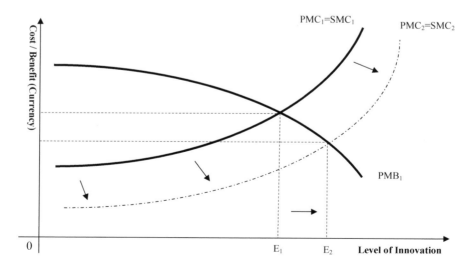

Fig. 3.3 The privately optimal levels of investment in innovation [19]

To proceed with the analysis, it is *assumed that higher levels of investment implies a stage beyond 'no innovation' and reflects any one of the higher types of innovative activity (adoption, adaption or creation), with the given technology* (For example, in order for a firm to progress from merely adopting technology to a higher level of activity like adapting the, it may have increased its level of spending on innovation). Based on this assumption, reductions in PMC and/or increases in the PMB will not only increase the level of innovation *but also* result in the firm engaging in higher types of innovative activities.

The second set of enablers in the conceptual model encourages innovation by lowering the cost of innovation. Swann [20] identified several types of cost situations associated with innovation and three cases are directly relevant for developing the ideas associated with this framework (Swan dealt specifically with process innovation. His ideas are extended to the general case of innovation here).

a. A Reduction in Fixed Costs, With no Change in PMC

Marginal cost is independent of fixed cost because fixed cost does not vary with additional innovation generated. However, the marginal cost will vary with variable costs associated with additional levels of innovation.

A reduction of fixed costs with no change in the PMC can occur, for example, if *initial technology* is gained cheaply through collaboration with outside research institutions or through links with MNCs or as affiliates of them; the recipient firm avoids investing in costly R&D infrastructure in-house. This, in turn, lowers the average cost of innovation and may encourage the firm to *initiate* in-house innovation. However, based on the assumption made earlier, it difficult to predict *a priori* the different types of innovation that might be encouraged by access to different modes of technology.

The framework therefore suggests that there will be a positive correlation between levels of innovation and the various modes of access to technology but cannot predict precisely the types of innovation they will generate.

b. An Increase in Fixed Costs, With a Reduction PMC

This case would capture instances such as purchasing technology, investments in new and upgraded technologies, setting up of dedicated R&D infrastructure and subcontracting out R&D. While such new spending will increase the fixed cost associated with innovation it will help to reduce the PMC of generating additional innovative activities. In the same vein, when firms hire professional or skilled workers, the quasi-fixed cost associated with such labour will increase but may lower the PMC associated with subsequent expenditures on innovation. (The flip side of this is having a large work force of unskilled workers which will increase the cost of undertaking innovation and therefore inhibit it.)

The prediction from the framework is that making royalty payments, having R&D infrastructure, subcontracting out R&D, and having a large share of professionals in the workforce will all be positively associated with higher levels of innovation, although the types of innovation that will be fostered cannot be predicted in advance.

c. **A Reduction in PMC, With no Change in (or Possibly Lower) Fixed Costs**

This case would reflect instances where the costs of inputs that go into innovation are reduced; this will lower the PMC and encourage more investment in innovation and (based on our assumption) increase the level of innovative activities engaged in by the firm. For example, collaborative research with outside institutions, receiving technical assistance and support from outside institutions, gaining new know-how from parent establishments or via supplier links with multinationals may lower the input costs associated with innovation, once innovation is initiated. This may occur explicitly (by reducing out-of-pocket expenses) and/or implicitly (by encouraging the absorption of expertise through the learning-by-doing process). Similarly, cluster-specific advantages may also lower the marginal cost of innovation.

The framework predicts that collaborative research with outside institutions, receiving technical assistance and support from outside institutions, gaining new know-how from parent establishments or via supplier links with multinationals and location in clusters are likely to encourage higher levels of innovation without providing additional insights on the types of innovation that might be encouraged.

Note that Fig. 3.3 cannot show the first case (a) that describes a situation where only the fixed cost changes with no effect on the marginal cost. It can only illustrate the effects of the change in the marginal cost of innovation; case (b), for example, refers to a situation where the fixed cost increases but the PMC of innovation is lowered. This is shown as a fall in the PMC from its original position (PMC_1) to a new lower position (PMC_2). Assuming the technology in place has not changed, the reduction in PMC will increase the equilibrium level of investment from E_1 to E_2. Based on the assumption made earlier, this could also suggest that the firm is undertaking higher types of activity. If on the other hand, the increase in fixed cost was due to newer technology being installed, there will be an upward shift of the PMB to indicate a higher stream of expected benefits from the new technology (not shown in the Fig. 3.3) and a downward shift of the PMC. This would suggest both greater investment in innovation and a move to a higher level of innovative activity.

3.8   Policy Environment

The final set of enablers, government policies, if carefully designed, also help subsidize the cost of acquiring or investing in technology and lowers the PMC of investments in innovation (with a given technology in place) as noted by the PwC Report referred to earlier. This can be seen from Fig. 3.4.

Once again, assume the firm achieves its private equilibrium by setting $PMB_1 = PMC_1$ and undertaking the corresponding level of investment in innovation shown by E_1. A government subsidy lowers the PMC from PMC_1 to PMC_2, thereby providing the incentive to increase the investment from E_1 to E_2. However, E_2 only represents the privately optimal investment and ignores the positive externalities generated by innovation, which is only captured by the higher SMB_1 and not PMB_1.

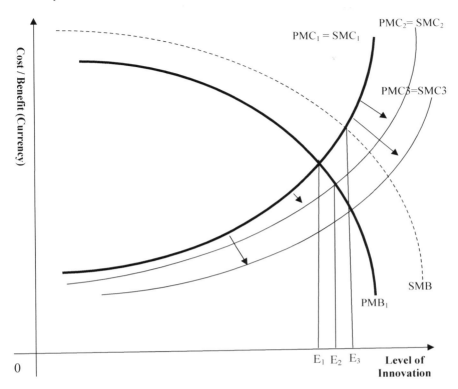

Fig. 3.4 The effect of a government subsidy on investment in innovation [19]

To ensure that investment is socially optimal, the firm must invest at the level indicated by E_3, which corresponds to the point where $SMB_1=PMC_1=SMC_1$ (The PMC is assumed to also reflect the SMC (Social Marginal Cost) as there are no externalities on the cost side). In order to provide an incentive to the firm to do this, it must theoretically be given a subsidy large enough to lower PMC_1 to PMC_3 so that it is equated to PMB_1 exactly at the level corresponding to E3. In practice, it is difficult to estimate the precise amount of subsidy required to achieve this result although government subsidy can be used to get firms to invest closer to the socially optimal level. The framework therefore predicts that incentives for innovation will encourage higher levels of innovation.

3.9 The Benefits of Innovation

The position of the PMB is determined by the technology in place; therefore, any development that requires new technology will shift the PMB curve upwards (to the right). The additional benefit (PMB) received from an additional ringgit of spending on investment, on the other hand, will represent a movement along the PMB curve. Several scenarios involve the PMB curve.

The later variations of the Linear Model and the Chain Linked Model suggest that firm-level innovation is driven by the additional net benefits (additional profits) to be reaped from meeting market demands [21, 22]. But this is just one possibility; another might be the need to compete in new and expanded markets or to lead the market by creating new or improved products. On the other hand, even if the firm did not have any of these ambitions, the mere fact that it faces increasing competition from others will force it to innovate, either in-house or through outsourcing.

Whether or not each of these cases will result in an upward shift of the PMB or will represent a mere movement along the PMB curve will depend on how the firm reacts to each of these situations. If the situation requires higher levels of investment expenditure with the technology currently in place, it would represent a movement along a given PMB and result in investments in higher levels of innovation. In contrast, if the situation calls for an investment in a new and better technology, it would shift the PMB upwards and result in more sophisticated technology *and* higher levels of innovation.

The main predictions from the analytical framework with regard to the enablers that lower innovation costs (These were incorporated in the conceptual framework) may be summarised as follows:

i.   The decision to innovate will be positive associated with royalty payments.
ii. Higher levels of innovation will be positively correlated with royalty payments, having R&D infrastructure, subcontracting out R&D, and having a large share of professionals in the workforce.
iii. Higher levels of innovation will be positively associated with collaborative research with outside institutions, receiving technical assistance and support from outside institutions, gaining new know-how from parent establishments or via supplier links with multinationals and location in clusters.
iv. Higher levels of innovation will be positively associated with incentives for innovation.

While suggesting that these factors will encourage higher levels of innovation (that is, go beyond the stage of 'no-innovation) the framework offers no direct predictions regarding the type of innovative activities (adoption, adaption or creation) these factors are likely to encourage. This will be investigated empirically in further sections.

3.10 Classifying the Types of Innovating Firms

The review of literature highlighted the existence of various types of innovations, their definitions and some practical problems in implementing them empirically. For the purpose of this study, innovation was defined as new or significant changes to goods or services; production or delivery methods. It is recognized that marketing and organizational innovations are also important but the data used in the study preclude their inclusion. As noted previously, the definition also encompasses aspects of both product and process innovation though the separate analysis should be performed to enable us distinguishing the two clearly. The data also do not distinguish between incremental and radical innovation. It was to circumvent these difficulties, that study adopted a classification used by the World Bank [23–25] and which was also consistent with the survey data generated.

The World Bank [23–25], in discussing the results of the *Productivity and Investment Climate Surveys* they conducted in collaboration with the Economic Planning Unit (EPU) of the Prime Minister's Department and the Department of Statistics (DOS) in Malaysia, introduced a new classification scheme that divided firms into four groups based on whether or not they were engaged in innovation and the type of innovation they were primarily involved in. The classification was as follows:

a. Non-innovator

A non-innovator is a firm that reported no innovative activity over the past two years (from the date of the survey).

b. Adopter

An adopter is a firm that built or improved its innovative capability. Adoption or *adoptive innovation* if it had upgraded its machinery and equipment or introduced new technology that had substantially changed the way its main product was being produced in the last two years, but did not include any of the other activities listed below.

c. Adapter

The adapter is a firm engaged in *adaptive innovation*. An adapter is any firm that had either upgraded an existing product line or entered new markets due to process or product improvements in quality or cost, or developed a major new product line, but had not filed any patents in the last two years.

d. Creator

A creator is a firm doing *creative innovation*, which is indicated if it had filed patents/utility models or copyright protected materials over the last two years.

Some limitations of these classifications must be noted. First, they may overlap; for example, an adapter may also be engaged in some adoption activities, or a creator may also be doing adoptive and adaptive work. But a firm in a given category cannot be engaged in activities associated with a higher category. Therefore, firms were classified as belonging to one of the four groups based on either no innovative activity or the type of innovative activity, they were *primarily* engaged in. Second, the assumption behind this classification is that successively *higher levels* of innovative capabilities are necessary for a firm to move from no innovation to adoption, adaptation and, finally, creation. But the classifications do not suggest anything about the *sophistication* of the technology underlying these three types of innovation. For example, an adapter in one subsector may be working with a more sophisticated technology than an adapter or a creator in another. The survey data did not have any information regarding the sophistication of the technology installed in firms. The mentioned classification is difficult to capture in the statistical analysis since future enterprise surveys conducted after 2007 did not included the same types of questions in the enterprise survey. Further classification of World Bank enterprise surveys divided the firms based on product and process innovation and information regarding filing patent or copy right application were dropped from the questionnaire. Finally, firms could be involved in product or process innovation or both but the firms were not required in the survey to distinguish between them. The strength of this classification is that it captures the innovative capabilities of firms without the need to distinguish between product and process innovation or radical and incremental innovation.

3.11 Conclusion

We reviewed the literature to identify the definition of innovation that will be used in the current study. The models of innovation were also reviewed and showed that they were not very useful in understanding factors that motivate the firm-level innovation. A conceptual model was therefore developed to identify the role of the various correlates of innovation discussed in empirical studies. Finally, a cost-benefit framework was used to understand how these correlates might encourage investment in innovation. Further we described the definitions of the types of innovation, the data, and the estimation procedure that was used commonly. This study also grouped the main correlates of innovation discussed in literature and explained how they were measured in the econometric models.

References

1. H.L. Sussman, *Victorian Technology: Invention, Innovation, and the Rise of the Machine* (ABC-CLIO, 2009)
2. J.A. Schumpeter, *Capitalism, Socialism and Democracy*, 3rd edn., vol. 3. 1942, (Routledge, New York, United State of America, 1975 Reprint), p. 249
3. R. Solow, *Robert M. Solow* (1990), p. 268–284
4. O. Gassmann, Opening up the innovation process: towards an agenda. R&D Manag. **36**(3), 223–228 (2006)
5. C. Freeman, *Technology Policy and Economic Performance: Lessons from Japan*, 2nd edn., vol. 2 (London, New York, 1987), p. 155
6. B.H. Hall, F. Lotti, J. Mairesse, Innovation and productivity in SMEs: empirical evidence for Italy. Small Bus. Econ. **33**(1), 13–33 (2009)
7. B.H. Hall, *Innovation and Productivity* (National Bureau of Economic Research, 2011)

8. B.H. Hall, J. Mairesse, P. Mohnen, Chapter 24—Measuring the returns to R&D, in *Handbook of the Economics of Innovation*, ed. by B.H. Hall, N. Rosenberg (North-Holland, 2010), pp 1033–1082

9. A.M. Knott, C. Vieregger, Reconciling the firm size and innovation puzzle. Organ. Sci. **31**(2), 477–488 (2020)

10. F. Bogliacino et al., Innovation and development: the evidence from innovation surveys. Lat. Am. Bus. Rev. **13**(3), 219–261 (2012)

11. S.M.P. Hosseini, S. Narayanan, Adoption, adaptive innovation, and creative innovation among SMEs in Malaysian manufacturing. Asian Econ. Pap. **13**(2), 32–58 (2014)

12. P. Vahter, J.H. Love, S. Roper, Openness and innovation performance: are small firms different? Ind. Innov. **21**(7–8), 553–573 (2014)

13. S. Lall, Competitiveness indices and developing countries: an economic evaluation of the global competitiveness report. World Dev. **29**(9), 1501–1525 (2001)

14. T. Xia, X. Liu, Foreign competition, domestic competition and innovation in Chinese private high-tech new ventures. J. Int. Bus. Stud. **48**(6), 716–739 (2017)

15. Matsumoto, M., et al., *Sustainability Through Innovation in Product Life Cycle Design* (Springer, 2017)

16. UNDP, *Malaysia Small and Medium Enterprises: Building and Enabling Environment* (United Nations Development programme, Malaysia, 2007)

17. R.R. Nelson, *National Innovation Systems: A Comparative Study*, vol. 8, no. 2 (Oxford University Press, 1993), p. 45

18. R. Rasiah, *Institutions and Public-Private Partnerships: Learning and Innovation in Electronics Firms in Penang, Johor and Batam-Karawang* (Institutions and Economies, 2017), pp 206–233

19. S.M.P. Hosseini, *A Study of Firm Level Innovation in Malaysian Manufacturing* (Universiti Sains Malaysia, 2015)

20. G.M.P. Swann, *The Economics of Innovation: An Introduction* (Edward Elgar Publishing Limited, UK, 2009), pp. 216–223

21. X. Fang, N.R. Paez, B. Zeng, The nonlinear effects of firm size on innovation: an empirical investigation. Econ. Innov. New Technol., 1–18 (2019)

22. M. Balconi, S. Brusoni, L. Orsenigo, In defence of the linear model: an essay. Res. Policy **39**(1), 1–13 (2010)

23. T.W. Bank, *Innovation Policy: A Guide for Developing Countries* (World Bank Publications, 2010)

24. World Bank, Malaysia firm competitiveness, investment climate, and growth, Report No. 26841-MA, in *Based on the Productivity and Investment Climate Survey conducted between December 2002 and May 2003 with reference period of 1999–2001* (Poverty Reduction Economic Management and Financial Sector Unit, East Asia and Pacific Region, 2005), p. 212

25. World Bank, Malaysia productivity and investment climate assessment update, Report No. 49137-MY, in *Based on the Productivity and Investment Climate Survey conducted between 2006 and 2007 with reference period of 2006*, ed. by J.W. Adams, et al. (World Bank Poverty Reduction and Economic Management Sector Unit, East Asia and Pacific Region, 2009), p. 260

Chapter 4
Machine Learning Approach to Identify Predictors in an Econometric Model of Innovation

Abstract Two common methods in measuring cross sectional data of innovation will be discussed together with the short comings of these methods when dealing with large sample size. Further, we aim to demonstrate how machine learning application can help us selecting the best appropriate exploratory variables. We elucidate several machine learning applications for predicting the best independent variables. Further implication of Probit and Ordered Probit models were compared with machine learning techniques, by using the most common variables in the literature to analyse the firm level of innovation.

Keywords LASSO · Regularization · Probit · Machine learning · Classification · Regression tree

4.1 Introduction

Industrializations involved computers that generate massive data regarding economic activities. Conventional statistical tests as well as econometric techniques such as linear regression are common tools for measuring determinant of innovation. Alternatively, the larger the data set requires more sophisticated tool of analysis. The selection of the explanatories for a regression test of innovation is often either arbitrary choice of researcher or is accordance to the theoretical background drawn from literature review. Further larger data sets are associated with bigger degree of freedom in the data set, which translates to more representable, more realistic estimation results and of course more flexible relationship between parameters. Machine learning techniques may allow for more effective ways to model complex relationships. This section describing few machine learning techniques that might predict variables requiring to estimate firm's decision to innovate or predictors for level of innovation.

As human and computers are generating more detailed data everyday several number of observations and several numbers of variables are generated. Extracting meaningful results and sorting and coding of data of this size is virtually impossible with conventional method of analysis. The challenge of dealing with massive datasets led to the development of big data and machine learning techniques. According to Varian, [1] data analysis in statistics and econometrics can be broken down into four categories: (1) prediction, (2) summarization, (3) estimation, and (4) hypothesis testing.

Machine learning deals with finding useful information or interesting pattern from a massive data set. Econometricians, statisticians, and data mining specialists are often looking for developing high-performance predictions for knowledge extraction, predictions of behaviour based on information discovery and information harvesting, data archaeology, data pattern processing, and exploratory data analysis. Machine learning tools offering ways that can usefully summarize various sorts of nonlinear relationships in the data. In the following g section, we will demonstrate examples of predictions.

4.1.1 Regression Based Analysis

To demonstrate how machine learning approach can be beneficial in predicting estimator of innovation we start with a simple regression. We demonstrate the conventional method of the regression then we will demonstrate how machine learning approach can help us identify the independent variables. Simple regression model consists of a dependent variable (y) as the predictor of several independent variables $y = f(x_1, x_2, x_3, \ldots x_n)$. In prediction of dependent variable, we can use the mean or median of the conditional distribution [1, 2]. Machine learning approach also using the same terminology as the independent variables are usually called "predictors" or "features." Machine learning aims is to find more suitable independent variables and eliminates irrelevant independent variables from an enormous list of independent variables to provide a better prediction of y as a function of x. Basic assumption of cross-sectional data is the sample data is naturally distributed. We typically looking for a best linear unbiased estimation "BLUE" with minimum sum of square residuals. Ideally, there might be better options of estimation especially when dealing with huge sample size and many variables. As Varian [2] mentioned some of machine learning approaches include nonlinear methods such as (1) classification and regression trees (CART); (2) random forests; and (3) penalized regression such as LASSO, LARS, and elastic nets. The advantage of LASSO over other estimations is the simplicity of the method and to predict which variable is not suitable for the regression.

Assuming the dependent or outcome variable that measures level of innovation in the study is categorical in nature. Two suitable models in this case are the simple Probit model to test whether a firm is innovating or not, and an Ordered Probit model that allows the dependent variable (type of innovation) to be ordered in an ascending manner beginning from no innovation and moving to successively higher levels of activity indicated by adoption, adaption and, finally creation.

As Menard [3] notes, when the dependent (or outcome) variables are measured on an ordinal scale, the usual methods of analysis have limitations. For instance, the Ordinary Least Squares (OLS) regression technique is problematic because the assumptions of OLS are violated when it is used with one or more discrete outcome variables. The Multinomial Logit or Multinomial Probit models would be better alternatives. However, when dealing with ordered information, the ordering in the outcome variables will be lost if the multinomial approach is used. The way to overcome this problem is to apply the Ordered Logit or Ordered Probit techniques.

The difference between the two is the different assumptions underlying them. As Gujarati ([4]: 594–595) noted, the cumulative distribution function (CDF) for qualitative response models can either be represented by the logistic or normal. The former gives rise to the Logit model and the latter to the Probit model. Gujarati ([4]: 614–615) goes on to say "that the in most applications the models are quite similar, the main difference being that the logistic distribution has slightly fatter tails…. That is to say, the conditional probability P_i approaches zero or one at a slower rate in Logit than in Probit". In most instances the models yield estimates that are similar, though the interpretations of the coefficients are different.

Many of the studies reviewed applied probability regression models such as Logit, Probit and Tobit but not the more advanced ordered models. Since innovation is relative concept that identified by product innovation, process innovation, organizational innovation, marketing innovation, in terms of related theories they can divide in to radical innovation, incremental innovation. To empirically demonstrate the theoretical framework each econometric test will be explained in detail and regression-based analysis will be reported.

4.2 Probit Model

The simple Probit model can be used to address the initial issue of the factors that predict whether or not a firm innovates. In this model, the outcome variable is binary in nature, taking a value of zero or one, as shown below (See [5] for details):

$$y = \begin{cases} 0 \text{ if a firm did not report any type of innovation} \\ 1 \text{ if a firm reported any type of innovation} \end{cases}$$

The Probit model estimates the probability that $y = 1$ as a function of the independent variables.

$$P = pr[y = 1|X] = F(X'\beta) \tag{1}$$

The advantage of Probit model over normal linear regression is that the predicted probabilities will be limited between zero and one. For the Probit model, $F(X'\beta)$ is the cumulative distribution function of the standard normal distribution.

$$F(X'\beta) = \Phi(X'\beta) = \int\limits_{-\infty}^{x'\beta} \Phi(z)dz \tag{2}$$

The Probit model is estimated using the maximum likelihood method.

In the Probit model, it is legitimate to interpret the *signs* of the estimated coefficients of the variables but the *magnitudes* of the coefficients should not be interpreted directly nor be compared with the coefficients from a normal linear regression. The estimated coefficients for the Probit model are 2.5 times bigger than those derived from a liner regression; in fact, since $\beta_{probit} = 2.5\beta_{OLS}$, comparisons or direct interpretation of the coefficient will be misleading. To remedy this problem, the marginal effects of the estimated coefficients in a Probit model are usually reported and discussed.

The marginal effect of a predictor is defined as the partial derivative of the event probability with respect to the predictor of interest. A more interpretation is the change in predicted probability for a unit change in the predictor.

Bartus [6] discusses the two methods of estimating marginal effects the average marginal effects (AME) and the marginal effects at the mean (MEM). The AME method involves computing the average of the discrete or partial changes over all observations, to yield the average marginal effects while the MEM approach computes the marginal effects at fixed values of the independent variables with the sample means being the most commonly used values. The literature is not conclusive about which might be the preferred approach. As Bartus ([6]: 309–310) points out, the main argument in support of the AME is that it is more realistic since the calculation of MEM might refer to either non-existent or inherently nonsensical observations, especially when dummies are among the regressors. He cites Green [5] as noting that while AME was more widely used, MEM, is an asymptotically valid approximation of AME (For a detailed technical exposition on how these two methods are derived, see [6]). In this study, the AME approach was used to compute the marginal effects.

4.3 Ordered Probit Model

The second issue on the factors that drive and enable different types of innovation was investigated with the aid of an Ordered Probit model. It has the advantage of generating directly the marginal effects of each covariate on the probability of engaging in each of the four types of activity. The cumulative probability $\Pr(y_i \leq j)$ is the probability that y is less than or equal to a particular value j (see [5, 7] for further details).

In general, the model can be specified as follows:

$$y_n^* = \beta x_n + \varepsilon_n \tag{3}$$

where

y_n^* is an exact but latent measure of the type of innovation engaged in by firm n
x_n a vector of explanatory variables likely to influence the type of innovation
β a vector of regression coefficients (parameters) to be estimated
ε_n a random error term assumed to be normally distributed and has a mean zero $N(0, 1)$.

The conditional probabilities of the ordered outcomes can be written in terms of cumulative probabilities as follows:

$$
\begin{aligned}
\text{Type of activity} &= \text{No innovation or 0 if } x\beta + \varepsilon \leq \mu_1 \\
&= \text{Adoption, or 1 if } \mu_1 < x\beta + \varepsilon \leq \mu_2 \\
&= \text{Adaption, or 2 if } \mu_2 < x\beta + \varepsilon \leq \mu_3 \\
&= \text{Creation or 3 if } \mu_3 < x\beta + \varepsilon
\end{aligned}
\tag{4}
$$

Note that μ_i's are thresholds that separate the types of innovation and are estimated along with the vector of parameters. For instance, the probability of engaging in creative innovation is

$$\Pr(creative) = F(\mu_3 - x\beta) - F(\mu_2 - x\beta) \tag{5}$$

where F(·) is a cumulative distribution function. The probabilities of engaging in other types of innovation are estimated similarly, with the use of proper threshold parameters. If the error term has a mean of zero then the probability F(·) is the standard normal cumulative distribution function underlying the Ordered Probit model.

4.4 The Expected Signs for the Explanatory Variables

Two models of estimation were explained. The first, a Probit model (Model 1), sought to identify the predictors of whether or not a firm will innovate. The outcome variable was categorical with a firm that reported any type of innovation during the reference period of survey data being assigned a value of 1, and zero, otherwise. All independent variables that would definitely suggest innovation were dropped as shown in the table below. The expected signs of the nine correlates of the innovation (not its level) are summarised (in the second column) in Table 4.1.

The second model (Model 2) tried to identify the drivers and enablers of the three different types of innovation (adoption, adaption and creation) and the characteristics associated with non-innovating firms using an Ordered Probit model. The question marks in Table 4.1 indicates uncertainty in regards to the estimation results.

Table 4.1 Expected signs of independent variables in model 1 and model 2

Variable name	Expected signs	
	Model 1	Model 2
Drivers of innovation		
Share of total sales exported	+	+
Market concentration (4-largest firm ratio)	?	
Food processing		?
Textiles		?
Garments		?
Wood and wood products		?
Chemicals and chemical products		?
Rubber and plastic		?
Machinery and equipment		?
Electrical machinery, apparatus, office, accounting and computing machines		?
Electronics (equipment and components)		?
Auto parts, motor vehicles and parts (reference category)		−
Furniture		?
Enablers of innovation		
Facilitators of innovation		
Small and medium firm size (<150 Workers) (reference category)	−	−
Large firm size (More than 150 Workers)	+	+
Equity ownership (extent of foreign equity participation)	+	+
Age of establishment	+	+
Highest level of education completed by CEO/Owner	+	+
CEO/Owner makes all its investment decisions independently	+	+
Lowers cost of innovation		
Sought collaboration in R&D from different sources		+
Staff exclusively for design/R&D		+
Technology transferred from parent establishment		+
Subcontracted out R&D		+
Received research/technology support from institutions		+
Supplier links to a MNC		+
Share of professionals and managerial workers in the workforce	+	+
Share of foreign workers in the workforce	−	−
Made royalty payments	+	+
Located in Penang	+	+
Policy environment		
Received government incentives for R&D		+

The complete list of independent variables used is shown in the Table 4.1. The theoretical framework by Hosseini and Narayanan [8] developed earlier only suggested a likely association (negative or positive as shown in the table) with higher types of innovative activities but it was incapable of suggesting which these factors are more likely to be associated with each of the different types of activities.

4.5 How Machine Learning Techniques Can Identify Predictor of Innovation

Data related to innovation in manufacturing can be analysed in different statistical packages. Since usually innovation data in manufacturing take place in national level large amount of data might be generated and will be available. Available firm level data requires exploratory data analysis along with consistency and data cleaning tasks.

In machine learning, independent variables are usually called "predictors" or "features" [2]. The focus of machine learning is to find some function that provides a good prediction of innovation activity as a function of predictor of innovation. Economist traditionally apply linear or logistic regression in these situations. One of the weaknesses of liner regression is inability to properly deal with the problems of non-linear and interactive effects of explanatory variables [9]. Both machine learning and data mining techniques can help us to solve the issue although the border line between the two concepts of data mining and machine learning are extremely narrow. Consequently, both processes employ the same critical algorithms for discovering data patterns. However, there might be better alternatives of simple regression, particularly when sample size increases. Few of these methods in machine learning for instance are: 1) classification and regression trees (CART); 2) random forests; and 3) penalized regression such as LASSO, LARS, and elastic nets.

They are two methods of machine learning first supervised approach and second unsupervised approach. Supervised approach, is functioning by human intervention in accordance to pre-defined category labels that are assigned to data based on the likelihood suggested by a training set of labelled data; and unsupervised approach, on the other hand do not require human intervention or labelled data at any point in the whole process [10]. We can apply supervised machine learning techniques to resolve the problem of identifying predictors (in our case innovation) by producing predictions of dependent variable from independent variables. The appeal of machine learning is that it manages to uncover generalizable patterns without overfitting the prediction. For instance, predict Y (the innovation of a firm) from its observed characteristics X based on a sample of n firms (Y_i, X_i). The algorithm would take a loss function $L(\hat{Y}, Y)$ as an input and search for a function $\hat{f}$ that has low expected prediction loss $E_{(y,x)}[L(\hat{f}(X), Y)]$ on a new data point from the same distribution.

Available data mining applications are:

- K-nearest neighbor classifiers (KNN) linear discriminant analysis (LDA) Classification

- Logistic regression (LR)

- Discriminant analysis (DA)

- Naive Bayesian classifier (NB)

- Artificial neural networks (ANNs)

- Classification trees (CTs)Classification trees (CTs)

The first step is to randomly divide the data into two groups, one for set of data for model training and the another set of data to validate or test the model. Error rates were often used as the measurement of classification accuracy of models [9]. Training data is to estimate a model, the validation data is to choose the model, and the testing data to evaluate how well the chosen model will performs (Often in reality validation and testing sets are combined).

K-fold cross-validation method can be applied to validate a suitable value for tuning parameter. Common choices for K are 10, 5, and the sample size minus 1 ("leave one out"). 10-fold and 5-fold cross validation is very common in machine learning techniques. For further detail please refer to [11].

4.5.1 Classification and Regression Trees

Let us assume if a manufacturing firm undertakes a process innovation activity then we code them as the outcome 1 and 0 otherwise. A discrete variable will be created. In machine learning it is known as a "*classification problem*" as [2] describes it. An example would be classifying manufacturing firms into "innovative firms" and "non-innovative firms" based on the characteristics of the firm. Decision trees can describe a sequence of decisions that results in outcome. This estimation approach commonly known as "classification and regression trees," or "CART". To present the use of tree models, STATA statistical software [12] were utilized to find a tree that predicts probability of process innovation in a manufacturing firm using just two dummy variables: being an export and having R&D investment in the firm. Data were obtained from micro-level World Bank database. The Enterprise Surveys are conducted and combined by the World Bank and its partners across all geographic regions and cover small, medium, and large companies. The data involves all economies following a global methodology for all indicators and it is available at (http://www.enterprisesu rveys.org).

Since the sample were divided into training and test sample in an arbitrary manner then each time, we run the test results may slightly diverge. In each time and for each section (training sample and testing sample) of the analysis a question arises for predictor of process innovation. The first node contains the value for everything that is true and second node contain everything else. For example, here we partition our sample based on whether the firm invests in research and development or not (In training sample and in testing sample). True values are firms who invested in R&D and false is everything else. Best situation occurs when uncertainty reduces the most. The aim of regression tree is to estimate how much uncertainty exist at each node. This is a metric between zero and one. Where lower values indicate less uncertainty and higher values means mixing at a node. The starting sets the percentage of impurity then by partitioning the data we calculate the uncertainty of the child nodes of that result.

The resulting tree is shown in Fig. 4.1, and the rules depicted in the tree are shown in Table 4.2. The text representation of tree can be explained as at node 1 if the firm having R&D investment <= 0.5 move to node 2 else move to node 3. At node 2 if the firm is an exporter <= 0.5 move to node 4 else move to node 5. At node 3 if the firm is an exporter <= 0.5 move to node 6 else move to node 7. The data consist of 149172 observations in which the sample were divided to learning sample of 74710 observations and test sample of 74462 observations.

In the above example for the simplicity of the explanations we demonstrated our results only for two independent variables. Furthermore, they were restricted to binary trees only (two branches at each node) although application of machine

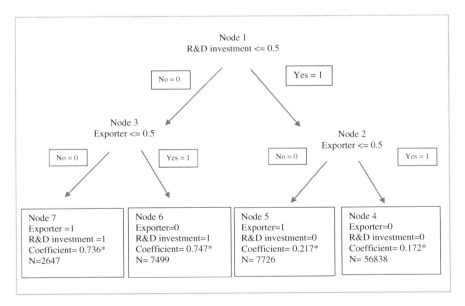

Fig. 4.1 Decision tree of the process innovation in manufacturing companies
Note Total number of the sample was 149172
* Represent significance of the coefficients in the boxes represent the coefficient of the regression

learning can be implemented for continuous dependent and independent variables and categorical variables as well. It turns out that there are computationally efficient ways to illustrate classification trees of this sort.

The above method of classification is very common for conditions and analysis of nonlinear as well as interaction variables. As explained in the previous sections same data can be estimated with a Logistic or Probit regression. To demonstrate let us estimate the probability of process innovation (Dependent variable) as a function of percentage of direct export, Amount of R&D expenditure (investment) results reported in Table 4.3. As Table 4.3 reveals research and development and being an

Table 4.2 Terminal node results for process innovation in manufacturing industries

Node	Node characteristics	Number of observations	Coefficient	Std. error	Z	P-value	95% confidence	Interval
4	Exporter=0 R&D investment=0	56838	0.173	0.002	108.530	0.000	0.169	0.176
5	Exporter=1 R&D investmet=0	7726	0.218	0.005	46.400	0.000	0.209	0.227
6	Exporter=0 R&D investment=1	7499	0.748	0.005	149.680	0.000	0.738	0.758
7	Exporter=1 R&D investment=1	2647	0.736	0.008	85.830	0.000	0.720	0.753
	Terminal nodes=4	Number of observations	R-squared	Average DV	Root MSE			
	Learning sample	74710	0.1996	0.255	0.390			
	Test sample	74462	0.1993	0.257	0.391			

Table 4.3 Result of Logit regression

Logit regression of new process innovation	Robust Coef.	Std. Err.	Odd ratio	Z	P-value	[95% conf.	Interval]
Exporter	0.188	0.019	1.207	9.890	0.000	0.151	0.226
R&D investment	2.579	0.018	13.183	144.700	0.000	2.544	2.614
Constant	-1.547	0.018	0.213	-202.050	0.000	-1.562	-1.532
Log pseudolikehood	-72001.191						
N	149,172						
Wald chi2(2)	21534.84						
Prob > chi2	00.000						
Pseudo R2	0.1518						

exporter firm are strong predictors of process innovation. Both the regression tree as well as logistic model suggests that having R&D investment and exporting are important predictors of process innovation.

4.5.2 Lasso Regression

Here we start by considering a standard multivariate regression model where we predict level of innovation in manufacturing as dependent variable (Y) as a linear function of a constant, b0, and P predictor variables. Further let us suppose that we have standardized all the (nonconstant) predictors with zero mean and variance one. Consider by minimizing the sum of squared residuals plus a penalty term of the form we select the coefficients $(b_1, \ldots, b_p)$ for these predictor variables [1].

$$\lambda \sum_{p=1}^{p} [(1 - \alpha)|b_p| + \alpha |b_p|^2] \tag{6}$$

The above model known as "elastic net regression" which also contains three other variations of the same model.

Ordinary least square (OLS) when $\lambda = 0$ (no penalty term)

Ridge regression when $\alpha = 1$ only quadratic constraint

Least absolute shrinkage selection operator (LASSO) when $\alpha = 0$

These are examples of regularization in a regression. The underlying theory of the mentioned methods are to shrink the coefficients of a regression equation to zero. In other words, regularization is a process to recognize parameters that are less effective in the model and select the most appropriate ones. What we explained here is just a fingertip of what machine learning techniques can do. Further analysis required for precise and robust estimation Table 4.4.

Table 4.4 Result of Lasso probit model

ID	Description	Lambda	Number of non-zero coefficient	Out of sample deviation ratio	CV mean deviance
1	First lambda	0.3285	0	0.0002	1.1378
78	Lambda before	0.0002	31	0.2056	0.9040
*79	Selected lambda	0.0002	31	0.2056	0.9041

Selection: 10 Folds Cross-validation

N = 149,172

Number of Covariates = 35

Note * lambda selected by cross-validation. Lambda selected based on stop () stopping criterion

4.6 Conclusion

We described the definitions of the types of innovation, and various types of methodologies dealing with innovation data, and the estimation procedure advantage and disadvantages of each one. It also grouped the main correlates of innovation discussed in Hosseini and Narayanan [8] framework and explained how they were measured in the econometric models that were in the study. The expected signs of the independent variables in both models were also indicated. Further we discussed how machine learning applications can be beneficial for analysis of firm level data.

References

1. H.R. Varian, *Replication data for: big data: new tricks for econometrics* (2019)
2. H. Varian, Big data: new tricks for econometrics. J. Econ. Perspect. **28**(2), 3–28 (2014)
3. S. Menard, *Applied Logistic Regression Analysis*, 2 edn., vol. 106 (Sage University Paper, United State of America, 2002), p. 111
4. D.N. Gujarati, *Basic Econometrics*, 4th edn, (Sangeetha, ed.) (McGraw-Hill, New York, 2003), p. 1034
5. W.H. Greene, *Econometric Analysis*, 5 edn., vol. 5 (R. Banister, ed.) (Prentice Hall, Pearson Education, New York, 2004), p. 1054
6. T. Bartus, Estimation of marginal effects using margeff. Stata J. **5**(3), 309–329 (2005)
7. D.A. Powers, Y. Xie, *Statistical Methods for Categorical Data Analysis*, 1 edn. (Emerald Group Publishing, United Kingdom, 2000), pp. 59–78
8. S.M.P. Hosseini, S. Narayanan, Adoption, adaptive innovation, and creative innovation among SMEs in Malaysian manufacturing. Asian Econ. Pap. **13**(2), 32–58 (2014)
9. I.C. Yeh, C.-h. Lien, The comparisons of data mining techniques for the predictive accuracy of probability of default of credit card clients. Expert Syst. Appl. **36**(2, Part 1), 2473–2480 (2009)
10. A. Ozgur, *Supervised and unsupervised machine learning techniques for text document categorization.* (Unpublished Master's Thesis, İstanbul: Boğaziçi University, 2004)
11. T. Hastie, R. Tibshirani, J. Friedman, *The elements of statistical learning: data mining, inference, and prediction.* (Springer Science & Business Media, 2009)
12. R. Mora Villarrubia, *CRTREES: stata module to compute classification and regression trees algorithms.* (2019)

Chapter 5
Big Data and Innovation; A Case Study on Firm Level Innovation in Manufacturing

Abstract This study investigated several aspects of firm level innovation in Malaysian manufacturing: the factors that influence the decision to invest in innovation activities; the extent of innovation; factors characterizing an innovating firm; the types of innovation and the factors that drive and enable them. Following the definition of Big Data, we drawn the data from a large representative survey from 2007 and 2015 of Malaysian manufacturing firms. The main findings unveil that while firm size, research and development investments, firms collaborative research, participation in international market through export among other indicators can positively influence firm level innovation. This section outlines the phases of the development of a coherent policy to foster, sustain and increase the level of innovation.

Keywords Manufacturing · Malaysia · National innovation survey · Innovation policy · Big data

5.1 Introduction

This section begins by describing the extent of firm level innovation as reported in several surveys in Malaysia. The first data collection of national innovation survey in Malaysia took place in 1994. Several survey data were also collected in the interval of four to five years. World Bank Productivity and Investment Climate Survey 2, 2007 was a collaborative work of department of statistics and World Bank. Recent data of Malaysian Enterprise Survey 2015 will enable us to compare the output of innovation through out those years. This helps to determine if firm level innovation growth is occurring and also to assess its prevalence in Malaysia and compares it with the findings of previous years.

Further we focus on the distribution of innovating and non-innovating firms by their characteristics such firm size, ownership, age of establishment, subsector of operation, modes of access to technology, locations of firms and other key variables. The qualitative analysis is deepened by a quantitative analysis of the factors that are

© The Author(s), under exclusive license to Springer Nature Singapore Pte Ltd. 2020
S. M. Parvin Hosseini and A. Azizi, *Big Data Approach to Firm Level Innovation in Manufacturing*,
SpringerBriefs in Applied Sciences and Technology,
https://doi.org/10.1007/978-981-15-6300-3_5

likely to distinguish an innovating firm from one that does not. For this purpose, a Probit model was estimated using the Big Data for large firms and the small and medium-sized firms (SMES), separately. The section closes with a discussion of the implications of the findings.

Economists historically require large amount of data for their analysis and lack of proper data limits reliable analysis. This is changing as new more-detailed data becomes available. There is a believe that "Big Data" is revolutionizing economic analysis [1]. The most common uses of "Big Data" by firms are for tracking innovation processes and innovation product outcomes, and for building a wide range of predictive models to predict innovation activities. We provide examples on how researchers may be able to measure determinants of innovation activities among firms by the help of "Big Data" to predict consequences of economic events or policies. A further but equally important issue arises in training economists to work with large data sets and the various programming and statistical tools that are commonly required for analysis. Big Data might be used to predict the risk involved for investing in innovation activities. From an empirical point of view, in order to analyse the level of innovation in manufacturing firms machine learning approach were used.

5.2 New Measures of Firm Level Innovation Activities

Department of statistics in collaboration with World Bank play an important role in tracking and monitoring enterprises innovation activity. Traditionally much of these efforts has been done using survey methods creating cross sectional data. For example, World Bank Enterprise Survey Statistics measures innovation activities by sending out surveys to firms worldwide. Collected data on the enterprises ease the availability of approximately hundreds of variables in a single survey. Sometime surveys are collected in longitudinal manner for various years. These data are aggregated into various innovation indices such as the product, process innovation. Other measures of innovation activities consist of firm size, firm ownerships, R&D expenditure, and technology transfer, etc. all these indicators rely on similar survey-based methodologies.

In the case of Malaysia combination of the two-survey data involve 1700 observations which can be consider a suitable example for a Big Data set to be encountered with level of innovation. The results of this analysis enable us to provide tangible pathways for future researchers aiming to integrate innovation in manufacturing and industrial research.

Big Data and Machine learning approach can provide a holistic picture of determinant of innovation and help us to understand that the available knowledge is greater than the sum of its parts [2]. Doornik and Hendry [3], distinguish three main types of "Big Data" according to these categories; First; Fat Data that is big cross-sectional dimension with small observations (N). Second; Tall Data that is big temporal dimension with small number of parameters, or Third; Huge that is big numbers of observation and big temporal dimension. Each classification is suitable for its own associate econometric method. One of the elements of big data is unstructured nature of big data, which imposes additional substantial challenges for proper econometric analysis. Conventional econometric methods of linear regression may not be suitable for variable selection. To elaborate our statement a case study example of Malaysian manufacturing firm as big data sample will be used after defining and discussion related to innovation. Increasing the pace of innovation promote greater R&D, attracts foreign knowledge-intensive companies more and promotes technology diffusion and acquisition of technologies.

5.3 Factors Affecting the Decision to Innovate in Malaysian Context

Innovation can be the main driver for economic growth. Participation in innovation activities inevitably involve risk taking. Further innovation is associated with substantial financial investments for firms. Firms constantly seeking to invest where their investment quickly pays off or generates business values. Innovation not only hard to define but also is intangible in terms of measurement and that reason increases the risk of undertaking innovation even more. A key strategy to address the risk is to use a highly interpretable innovation model that decision makers can understand and trust [4]. The main form of innovation activities in Malaysia is the adoption of technologies and innovation from highly industrialized countries [5, 6]. While a larger proportion of small and medium enterprises in Malaysia doing adaption of innovation were improving their quality and cost, a marginally bigger share of large firms are upgrading their product line entirely.

5.4 Extent of Firm Level Innovation

According to World Bank enterprise survey in 2015, in Malaysia around 70 (12%) out of 585 manufacturing firms were engaged in product innovation and 304 (52%) out of 585 of firms were engaged in process innovation. Considering product and process innovation 250 (42.74%) of firms only reported one of the types of innovation (either product or process innovation). Further 62 (10.6%) of firms reported both product and process innovation in 2015. Although when comparing the older sample data of 2007 that was collected from investment climate survey 715 out of the sample of 1115 firms were engaged in some form of innovation in 2006. Thus, 64.1% of all firms in the manufacturing sector were innovating firms in 2006. While the data are not strictly comparable due to differences in the definition of innovation, sampling methods and coverage, this figure is much higher than the figures reported in the four National Innovation Surveys that covered the periods from 1997–2008 (Table 5.1).

Table 5.1 Extent of innovation

Survey years	Sample size (stage two)	Non-innovating firms		Innovating firms	
		Number	Percentage	Number	Percentage
1994 (NSI-1)	412[a]	144	35	268	65
1997–1999 (NSI-2)	1044	825	79	219	21
2000–2001 (NSI-3)	749	487	65	263	35
2002–2004 (NSI-4)	485*	223	46	262	54
2005–2008 (NSI-5)	1212*	588	49	624	51
2006 (PICS-2)	**1115**	**400**	**36**	**715**	**64**
2015 (MES)	**585**	**273**	**47**	**312**	**53**

Estimated based on Productivity and Investment Climate Survey 2, 2007 and Malaysia Enterprise Survey 2015 [7, 8]
Note * Includes firms in manufacturing and services sector

It is also important to note the formation of panel data from the two survey data sets were impossible due to anomaly of the variable definitions. Only the first National Innovation Survey (1994) that covered a smaller sample of 412 firms reported a higher figure of 65%, but as survey data where old and definitions are varied, the results of the first survey must be treated with caution.

The services sector is currently the dominant sector in the Malaysian economy. By the end of the 9th Plan period (2006–2010), services had recorded the fastest annual growth rate (of 6.8%) and had raised its share of the GDP to 58% [9]. A recent study of innovation in the services sector, using comparable data and definitions utilized in this study, reported that 49.5% of service firms were innovating [6]. This suggests that larger proportion of manufacturing firms were active in innovation, even when compared to the dominant services sector.

Despite the fact that these figures are not strictly comparable, it is safe to conclude that the number of firms involved in innovation in the manufacturing sector has seen a rising trend over the years. This is not surprising because the Malaysian government has begun to pay more attention to policies and incentives that encourage innovation, especially in the manufacturing sector, in recent years. The extent of innovation varied by firm size; while only 47% of all small firms were innovating, the figure increased to 73% for medium sized firms and to 85.5% for large firms. The proportion of small firms that was innovating was therefore below the manufacturing average.

5.5 Characteristics of Innovating and Non-innovating Firms

The distribution of innovating and non-innovating firms by characteristics such firm size, ownership, age of establishment, subsector of operation, modes of access to technology, locations of firms and other key variables are given below. They suggest that while both groups have some characteristics that distinguish them from one another, they also share some similar characteristics.

5.5.1 *Innovation by Firm Size*

A somewhat different picture emerges when the firms are divided into innovators and non-innovators. Table 5.2 indicates that innovating firms were fairly equally divided across all firm sizes during 2007 and 2015. However, in the case of non-innovators, small firms were predominated during 2007 and larger firms were predominated in 2015. Almost 67% of non-innovators were small firms in 2007 and 26.37% were non innovators in 2015 while medium sized forms accounted for another 21.3% in 2007 and 50.18% in 2015 survey. Only 12.3% of the large firms were not engaged in innovation in 2007 however the number has increased in 2015 to about 22.3% non-innovators. About 35% of innovators were large firms in 2007 survey although 55.13% were large and innovators in 2015 survey. And although the majority of firms in 2007 sample were small, the proportion of large innovating firms marginally exceeded the proportion of small innovating firms in that year. It can be observed that innovating firms are getting larger in size from 2007 to 2015.

Table 5.2 Innovators and non-innovators by firm size

Firm size	Non-innovators	Non-innovators	Innovators	Innovators	Total	Total
	2007	2015	2007	2015	2007	2015
	%	%	%	%	%	%
Small sized enterprises (5–50 employees)	66.50	26.37	33.01	19.23	45.02	22.56
Medium sized enterprises (51–150 Employees)	21.25	50.18	32.31	25.64	28.34	37.09
Large sized enterprises (more than 151 employees)	12.25	22.34	34.69	55.13	26.64	39.83
Total	100.00	100.00	100.00	100.00	100.00	100.00

Estimated based on Productivity and Investment Climate Survey 2, 2007 and Malaysia Enterprise Survey 2015 [8]

Table 5.3 Innovation by ownership

	Non-innovators	Non-innovators	Innovators	Innovators	Total	Total
	2007	2015	2007	2015	2007	2015
	%	%	%	%	%	%
Purely domestically owned firms	81.00	79.49	64.48	72.12	70.40	75.56
Foreign equity less than or equal 30% (joint venture)	3.50	11.36	4.90	7.05	4.39	9.05
Foreign equity more than 30%	15.50	9.16	30.63	20.83	25.20	15.38
Total	100.00	100.00	100.00	100.00	100.00	100.00

Estimated based on Productivity and Investment Climate Survey 2, 2007 and Malaysia Enterprise Survey 2015 [8]

5.5.2   Innovation by Ownership

It is encouraging to note that the bulk of the innovating firms were domestically owned. They accounted for about 65.5% of innovators. Foreign owned firms (with more than 30% equity held by foreigners) accounted for just one quarter of the innovators (Table 5.3). This seemingly contradicts the widely held view that foreign firms lead in innovation.

However, domestically owned firms comprised a larger majority (81%) of non-innovating establishments relative to innovators (64.5%). In comparison, foreign owned firms only accounted for about 15.5% of non-innovators relative to 25.2% of innovators.

5.5.3   Innovation by Age of Establishment

Table 5.4 indicates that in both 2007 and 2015 survey data older firms (operating for more than 15 years) were dominant among innovators. Younger firms (in operation for 10 years or less) comprised another 27.4% in 2007 and 11.21% in 2015. A similar pattern prevailed among non-innovators. Oldest and youngest firms also formed the majority here, although the concentration of the oldest firms was marginally lower and the concentration of the youngest firms was higher among non-innovators as compared to innovators.

Table 5.4 Innovation by age of establishment

	Non-innovators	Non-innovators	Innovators	Innovators	Total	Total
	2007	2015	2007	2015	2007	2015
	%	%	%	%	%	%
Age of the establishment less than and equal to 10 years	32.75	16.87	27.41	11.21	29.33	13.85
Age of the establishment 11 to 15 years	19.75	23.06	23.50	27.00	22.15	25.13
Age of the establishment more than 15 years	47.50	60.07	49.09	61.81	48.52	60.00
Total	100.00	100.00	100.00	100.00	100.00	100.00

Estimated based on Productivity and Investment Climate Survey 2, 2007 and Malaysia Enterprise Survey 2015 [8]

5.5.4 Innovation by Subsector

By subsector, the comparison of the two years did not show any significant changes from 2007 and 2015 survey. In both years an interesting pattern emerges (Table 5.5); two subsectors, rubber and plastics and food processing had the largest concentration of both innovating and non-innovators. However, in comparing the two, it should be noted that while rubber and plastics had the highest share of innovators, food processing had the highest share of non-innovators. It will be noted that these were also the least concentrated subsectors.

In terms of innovative firms, the sub-sectoral innovation did not significantly change since 2007 where 21.3% were located within the rubber and plastics subsector in 2007 and 27.3% in 2015. This was followed by food processing that housed 20.3% of all innovators. The other key subsectors include chemical and chemical products, electronics and, finally, electrical machinery. Given the different levels of technology employed in these subsectors, the innovative activities can be expected to be very different as well.

The concentration of non-innovators was also high in food processing and rubber and plastics; these two subsectors together accounted for 46% of all non-innovators. Non-innovators were also concentrated in garments (12%), furniture (9.8%), and machinery and equipment (9%).

Table 5.5 Innovation by subsectors

	Non-Innovators	Non-Innovators	Innovators	Innovators	Total	Total
	2007	2015	2007	2015	2007	2015
	%	%	%	%	%	%
Food processing	24.50	32.60	20.28	22.12	21.80	27.01
Textiles	4.00	–	3.36	–	3.60	–
Garments	12.00	15.40	5.87	14.10	8.07	14.70
Wood and wood products	4.00	4.40	1.68	4.81	2.50	4.62
Chemicals and chemical products	4.50	15.40	8.95	15.38	7.35	15.37
Rubber and plastic	21.50	4.74	27.27	5.45	25.20	5.13
Machinery and equipment	9.00	3.66	7.97	3.85	8.34	4.10
Electrical machinery and apparatus and office, accounting, and computing machine	2.25	–	3.92	2.88	3.32	4.27
Electronics (equipment and components)	6.00	12.45	8.39	18.59	7.53	15.73
Auto parts motor vehicles and parts	2.50	–	3.50	1.60	3.14	1.03
Furniture	9.75	–	8.81	4.81	9.15	2.91
Others	–	11.35	–	6.41	–	5.13
Total	100.00	100.00	100.00	100.00	100.00	100.00

Estimated based on Productivity and Investment Climate Survey 2, 2007 and Malaysia Enterprise Survey 2015 [8]

5.5.5 *Mode of Access to Technology*

Innovating firms benefitted from access to technology or technological expertise through several modes as indicated in Table 5.6; collaborative research (Collaboration involved working with other firms, universities, multilateral agencies, or research institutions to develop or raise the innovative capability), obtaining technology from parent plants and through supplier links with multinationals were the most commonly utilized modes of access. Although both questionnaires in 2007 and 2015 did not directly ask questions related to technology transfer or collaborative research, we can observe related information through indirect respond to other questions. For

Table 5.6 Modes of Access to Technology

	Non-Innovators	Non-Innovators	Innovators	Innovators	Total	Total
	2007	2015	2007	2015	2007	2015
	%	%	%	%	%	%
Sought collaboration in R&D from different sources	2.50	2.43	58.04	21.21	38.12	11.97
Staff exclusively for design/R&D	2.50	1.39	16.78	23.91	11.66	12.82
Technology transferred from parent establishment	2.50	00.00	52.45	20.54	34.53	10.43
Subcontracted out R&D	2.75	4.86	8.81	42.76	6.64	24.10
Received research/technology support from institutions	0.25	2.08	16.50	0.67	10.67	1.37
Supplier to a MNC	1.00	6.60	23.64	43.77	15.52	25.47

Estimated based on Productivity and Investment Climate Survey 2, 2007 and Malaysia Enterprise Survey 2015 [8]

Note A firm can utilize several modes of access

instance, developing a new product line entirely through collaboration with research institutions is indication of receiving R&D support from other institution. Whereas 58% of all innovating firms had collaborated with others in undertaking R&D in 2007 the percentage fall to 20.54% of firms had collaborated in research in 2015, in 2007 around 52.5% gained technology from their parent establishments which reduced to 20.54% in 2015 and 24% accessed technology through their links as suppliers to multinationals in 2007 this value has sharply increased to 44% in the latest survey data (Note that a firm can utilize several modes of access to technology). About 17% in 2007 and 0.7% of firms in 2015 had received research or technological support from institutions such as SIRIM, a solution-provider in quality and technology; Malaysian Agricultural Research and Development Institute (MARDI) that does research in agriculture, food, and agro-based activities; and the Rubber Research Institute of Malaysia (RRIM),specializing in rubber and rubber-related products.

Finally, a small proportion of firms (8.8%) that reported innovation had actually outsourced their innovative activities this figure had sharp incline into 43% during 2015. These firms enjoy the fruits of innovation without engaging in innovation in-house. This phenomenon is noted in the literature as 'open innovation' [10]. The idea is that firms can go beyond the traditional concept of in-house innovation to leverage and gain from expertise outside the firm to benefit from technological advances.

5.5.6 Innovation by Other Key Variables

Based on Table 5.7, it is clear that proportionally more innovating firms were export oriented and they had younger (less market experience) university educated CEOs or owners. More innovators also bought their technology through royalty payment. The proportion of royalty payments had sharp increase in comparison between 2007 to 2015 survey. Interestingly, a larger proportion (45% in 2007 and 27.27% in 2015) of innovators also had a high share of foreign permanent workers in their workforce.

Non-innovators, on the other hand, had a higher than average share of professionals and a lower than average share of foreign workers in their workforce. They were also run by CEOs and owners who had the power to make decisions independently. Although the survey data in 2015 eliminated the question in regards to CEO education. It appears that neither a larger share of professional and managerial workers nor foreign workers were associated with innovation. The data also suggest that more experienced CEO/managers, independent and sole owners tend to shy away from innovation.

Table 5.7 Innovation by other key variables

Variables	Non-Innovators % 2007	Non-Innovators % 2015	Innovators % 2007	Innovators % 2015	Total % 2007	Total % 2015
CEO/owner makes all its investment decisions independently	16.00	5.90	9.93	7.07	12.11	6.50
Export orientated; (More than 10% of sales exported)	36.25	35.76	59.58	52.86	51.21	44.44
Made royalty payments	4.00	6.60	13.85	43.77	10.31	25.47
CEO/Owner is a university degree or higher degree holder	28.75	–	53.01	–	44.30	–
CEO/Owner years of experience in the market if the share is bigger than its mean of 12.52	–	44.44	–	30.64	–	37.44
Percent share of professionals and managerial workers if the share is bigger than its mean of 12.77 during 2007 and 13.34 during 2015	40.75	51.04	36.22	13.47	37.85	31.97
Percent share of foreign permanent workers if the share is bigger than its mean of 23.80 during 2007 and 9.03 during 2015	37.75	20.83	44.62	27.27	42.15	24.10

Estimated based on Productivity and Investment Climate Survey 2, 2007 and Malaysia Enterprise Survey 2015 [8]

5.6   Factors Motivating Innovation

Small firms accounted for 45% in 2007 survey and 22.56% in 2015 of all firms while medium-sized firms comprised another 28% in 2007 and 37.09% in 2015. Size emerged as a significant variable associated with innovation in the pooled data in the both surveys. Therefore, it seemed reasonable to analyse (run separate regressions) for large firms and small and medium sized firms (SMEs).

To ascertain if this was appropriate, the likelihood ratio test was run [11, 12]. First, four OLS regressions were estimated for each survey data: one for the pooled sample with all variables and two dummies to capture firm size (with small firms as the reference category), and one each for the segmented samples (small, medium and large). Maximum Likelihood (ML) estimates for the separate and pooled samples suggest that the hypothesis of equal regression (or slope) parameters across the firm size can be rejected (LR = 44.30, df = 30, p = 0.0447) for the former survey and (LR = 43.20, df = 28, p = 0.0327) for the latter survey. Therefore, it appeared appropriate to analyse the samples separately.

As the first step in the analysis, the factors that motivate innovation are examined, leaving aside for the moment the type of innovation. For this, a simple Probit model was estimated. The dependent variable was categorical in nature because it measured whether or not the firm was engaged in innovation (regardless of the type of innovation). In order to form the dependent variable of the Probit model, innovating firms were separated from non-innovating by joining the three categories of adoption, adaption, and creation or joining product and process innovation together as the innovating category and classifying the remaining firms as the non-innovators. An innovating firm was assigned a value of 1, while a non-innovator was given a value of zero.

Of the drivers and enablers discussed earlier, all variables that could reflect endogenous choices made by the firm that almost ensures there is innovation were dropped. These include variables that capture the various modes of access to technology except royalty payments. Purchasing technology (proxied by royalty payments) is a necessary but not sufficient condition for innovation. Also excluded were variables like having R&D staff and receiving incentives for R&D.

The independent variables that were retained in the model are shown in Table 5.8. To capture the impact of competition in driving innovation two proxies were used—percent of sales exported and degree of market concentration (measure by the 4-firm market share ratio (Dummies for the various subsectors were dropped due to high multicollinearity with the concentration ratio)). Firm level facilitators were captured by age of firm, size of firm, and percent of equity held by foreigners. Organizational effects were proxied by the Chief Executive Officers (CEOs) or owner's years of experience in the market, the education of the CEO and whether or not they were the sole decision makers. Question related to whether or not the CEO/owner has education degree education were eliminated in 2015 survey data and were replaced with CEO's years of market experience. Of the factors that lowered the cost of innovation only the percentage share of professional and managerial workers in the workforce (and its flip side, the share of foreign workers in the work force that was only available for 2007 survey data) and royalty payments were retained. Finally, to account for possible cluster-specific advantages, a dummy variable was used to separate Penang (Known as Silicon-Valley of Malaysia) from the other locations.

It will be recalled that the cost-benefit framework discussed in Hosseini and Narayanan [6] predicted that the percentage share of professional and managerial workers in the workforce, royalty payments and the location variable (capturing cluster-specific advantages) will all be positively associated with the decision to innovate.

Since the coefficients of the Probit model have no direct interpretation; therefore, the focus is primarily on the marginal effects and the full results of the Probit estimation were not reported here. The model was estimated using robust standard errors to address the minor Heteroskedasticity problem. The VIF and the Pearson correlation tests were done to rule out issues of Multicollinearity with the estimators. The results of these tests are is available upon request from the author. Both tests confirmed that we could not reject the null hypothesis of no Multicollinearity among the parameters.

The signs of the marginal effects are the same as the signs of the coefficients. Table 5.8 shows the marginal effects for the pooled sample as well as for the large firm and SME samples for both 2007 and 2015 survey data.

5.7 Pooled Sample

Dynamics of innovation activities by firms has changed during the time of the two surveys. Turning first to the pooled data, six factors for 2007 and four factors were significantly increased the probability of a firm innovating. Several changes are highlighted when comparing the result of the two-survey data. Firstly, market concentration and export are significant predictors of innovation during 2007 and are not significant compare to 2015 data. Although CEO's education is important predictor of innovation activity in 2007, experience of CEO/owner has no influence on the probability to innovate during 2015 survey. The strongest predictors of whether or not a firm innovates appears to be firm size in 2007 and royalty payment in 2015. The probability of innovating firms in 2007 increases by 24.7% for a large firm, relative to a small firm while the probability of innovating increases by 17.6% for a medium sized firm relative to a small firm. Medium firms in 2015 were not significantly innovative compare to small firms in 2015. Large firms in 2015 has 21.81% higher probability of innovation comparing with small and medium firms in the same year.

Consistent with the prediction of the theoretical framework Hosseini & Narayanan [6], firms paying royalties were more likely to be innovating; the probability of innovating increases by 20.3% in 2007 and 43.36% in 2015 for such firms as compared to firms not paying royalties.

The decision on whether or not to innovate was also strongly correlated with the educational qualification of the CEO or owner. Firms headed by CEOs or owners with university or higher degrees have 11.4% higher probability of engaging in

Table 5.8 Marginal effects of the predictors of innovation

	2007 survey data			2015 survey data		
	Pooled sample	SMEs	Large firms	Pooled sample	SMEs	Large firms
	dy/dx	dy/dx	dy/dx	dy/dx	dy/dx	dy/dx
Drivers of innovation						
Share of sales exported directly (%)	0.11***	0.17***	0.05	0.07	0.04	0.01
Market concentration (4-largest firm ratio)	−13.33*	−19.58**	−1.15	30.90	13.31	28.00
Enablers of innovation						
Facilitates innovation						
Medium sized firm (50–150 Workers)	17.60***	–	–	−6.59	–	–
Large firm size (more than 150 workers)	24.65***	–	–	21.81***	–	–
Age of establishment	−0.03	−0.19	0.52**	−0.08	−0.02	−0.03
Equity ownership (% foreign)	0.02	0.03	0.01	−0.02	−0.81**	0.18
University degree or higher degree completed by CEO/owner (in 2007) Years of CEO market experience (in 2015)	11.39***	14.28***	12.33***	−4.43	−3.97	−0.04
CEO/Owner makes all its investment decisions independently	−4.40	−7.49	−1.71	2.30	2.40	3.72
Lowers cost of innovation						
Share of professionals and managerial workers (%)	0.01	−0.23	0.20	−0.58***	−0.95*	−0.31***

(continued)

Table 5.8 (continued)

	2007 survey data			2015 survey data		
	Pooled sample	SMEs	Large firms	Pooled sample	SMEs	Large firms
	dy/dx	dy/dx	dy/dx	dy/dx	dy/dx	dy/dx
Made royalty payments	20.34***	22.78***	15.62**	43.36***	42.26***	34.92***
Share of foreign permanent workers (%)	0.02	0.10	−0.07	–	–	–
Penang	6.95***	8.50**	5.47	19.21***	1.11	26.31***

Note Coefficients are expressed in percentages
***indicate significant at 1% level, **indicate significant at 5% level, *indicate significant at 10% level

innovation as compared to firms led by those without tertiary education. On the other hand, although not significantly firms lead by old CEOs tend to be less risk adverse in 2015.

Being located in Penang was another factor that increased the probability of innovation; Penang based firms had an almost 7% in 2007 and 20% in 2015 higher probability of innovating, relative to firms located elsewhere. This result too was consistent with the prediction of the analytical framework.

During survey of 2007 the share of total sales exported was positively and significantly associated with engaging in innovation, though the impact was not large; a 10% increase in the share of total sales exported raised the probability of innovation by 1.1%. It is consistent with the view that increased competition drives innovation. However, during the survey in 2015 although the probability was positive the share of export and market concentration variables were not significant. In the literature there is no unanimity that export can increase competitiveness of the firms. Some argue that firms are induced to innovation activity just because they export to foreign markets with a strong demand growth (Fassio [13]. Possible reason for insignificant factor of export is because the erstwhile industries in Malaysia are still in assembly operation and component manufacturing without moving toward high value-added exportable products [14]. Further export destinations is indirectly influencing the demand side and can differ innovation activities a lot along several other dimensions (level of competition; geographical and cultural distance; institutional setting), and each of them even might induce different types of innovation outcomes [15]. Through interactions with foreign customers, firms who export benefit from international technological spillovers in the forms of know-how and new technological capabilities: hence, higher innovation and productivity is a consequence of exporting, rather than export led to innovation theory. Our result is in line with Woerter and Roper [16] where foreign demand for export is not sufficient to have strong impact on product or process innovation in the case of Irish firms.

On the other hand, during 2007 survey market concentration was significant at the 10% level but had a negative association with innovation. An increase in the concentration ratio of 0.1 lowered the probability of innovation by 13.3% during 2007 but has no significant effect found during 2015 survey data. Again, it is consistent with the argument that competition rather than monopoly power drives innovation.

Contrary to the predictions of the framework, neither the share of managerial and professional workers nor the opposite effect of a large share of immigrant workers featured significantly in the decision to innovate. In fact, during 2015 higher number of professional workers is associated with lowering the probability of innovation.

5.8 SMEs and Large Firms

Examining the factors that predicted innovation *within* each firm-size group was more revealing. In 2007 survey for both large firms and SMEs, royalty payments and the CEOs with tertiary education qualifications motivated innovation, although both factors had a stronger effect in the case of SMEs relative to large firms. In 2015 survey only royalty payments motivated innovation positively. Other factors such as share of foreign equity ownership reduced the likelihood of innovation among SMEs. Share of professional workers reduce the probability of innovation in both SMEs and large firms through the impact of among SMEs were stronger. Being located in Penang increase the probability of innovation among the larger firms.

During 2007 survey higher market concentration had a large dampening effect on innovation among SMEs but had no significant impact on large firms though the coefficient had a negative sign. An increase of 0.1 in the concentration ratio lowered the probability of innovation among SMEs by 19.6%. A more competitive market environment was therefore more important in motivating innovation among SMEs rather than large firms.

A similar message emerged when competition through participation in the export markets was examined in 2007 survey. Exposure to the export market was a strong driver of the decision to innovate among SMEs but did not have a similar impact on large firms. A 10% increase in the sales exported raised the probability of innovation among SMEs by 17%. Although the sign of these variables stayed positive during 2015 survey data the coefficients were not significant.

Larger and better-established firms during 2007 survey had a 0.5% higher probability of innovation relative to younger large firms but age had no significant impact among SMEs in the latter survey or other firms. If anything, younger SMEs were more likely to innovate as indicated by the negative sign associated with the coefficient during both two conducted surveys.

Finally, being located in Penang increased the probability of innovation among SMEs by 8.5% in 2007 but made no significant difference to large firms although during 2015 Penang had no significant difference to SMEs and increased the probability of innovation by 26.31 among large firms.

The high share of professionals and managers had no effect in motivating innovation among firms just as a large share of foreign (unskilled) workers did not appear as an obstacle to innovation in firms. While the coefficient was negative in the case of large firms, the magnitude of the effect was small and statistically insignificant during 2007. In the period of the two surveys the impact of the negative coefficients turned out to be stronger and more significant in 2015 survey in particular among SMEs. The high share of professional workers reduces the probability of innovation by 95% among SMEs whereas share of professional workers reduces the probability of innovation only by 0.31%.

5.9 Summary and Discussion

The qualitative analyses suggest the following characteristics of innovating firms. They were equally distributed across all firm sizes, were largely domestically owned and were firms that had been operation for more than 15 years. They were concentrated in, the rubber and plastics, food processing, chemical and chemical products and electronics subsectors and were located mainly in Johore, Selangor and Penang. Innovating firms also tended to be well established (older) firms. The majority of firms were also involved in exports. Interestingly, innovators had a larger than average share of foreign workers and a lower than average share of professional and managerial workers in their workforce.

More innovating firms were paying royalties and they have built up their technological expertise by relying on one or more of the following; collaboration in R&D with others, technology transferred from parent establishments, technological support from public institutions and expertise gained through their role as suppliers to multinationals. Finally, the majority of innovating firms were led by CEOs or owners who had tertiary qualifications or more market experience.

Non-innovating firms, on the other hand, were predominantly small in size, were overwhelmingly domestically owned and were either very old or very young. Like innovating firms, they too were concentrated in food processing and rubber and plastics and were located in the three main states of Johore, Selangor and Penang. Unlike innovators, however, they were also in subsectors such as garments, machinery and equipment and lower end electronic equipment. A smaller share of non-innovators was also engaged in exports. In striking contrast to innovators, non-innovators had a lower than average share of foreign workers and a larger than average share of professional and managerial workers in their workforce.

Not surprisingly, only four percent of non-innovators were paying royalties and less than three percent was involved in upgrading their technological expertise through collaboration or other means. A smaller proportion of non-innovators were led by CEOs or owners with tertiary qualifications or experienced CEOs while a larger proportion was led by owners who made investment decisions on their own.

The results of the Probit model supports the view that competition was an important driver of innovation but only in the case of SMEs during 2007 as indicated by the significance of export orientation and (lower) degree of market concentration. However, the significant impact of competition faded away by latter survey in 2015. These factors were not significant in the case of large firms, possibly because these firms have already acquired a significant market share and have established their hold on their market segment.

However, initial access to technology (proxied by royalty payments) and enlightened and pro-active CEOs and owners (proxied by their tertiary qualifications) played an important positive role in pushing a firm to innovate both in the case of large firms and SMEs. More experience CEOs on the other hand were less risk adverse and were not taking the risk of investing in innovation activities.

The locational advantages provided by Penang, the "Silicon Valley" of Malaysia was a key to motivating innovation among SMEs but had no significant impact on the large and well-established firms during 2007. The flipped situation happened by 2015 survey where SMEs in Penang were less innovative compare to larger firms.

Finally, a larger share of professional and managerial workers did not predict innovation while a large share of foreign workers in the firm did not inhibit innovation. The data also suggest that independent and sole owners are more risk averse and tend to shy away from innovation even when they have more market experience.

To summarise, a competitive environment (exposure to export markets and functioning in a competitive subsector), being located in Penang, being led by fresh to the market owners or CEOs with tertiary education and purchasing technology were significantly associated with innovating SMEs. In the case of large firms, only age, purchasing technology and being headed by younger CEOs or owners with tertiary education predicted innovation.

In closing, a preliminary assessment regarding the sophistication of the innovation occurring in firms is attempted, although the survey data do not permit such an exercise. It was noted that while most of the innovative activities were adaptive rather than creative, the proportion of SMEs and large firms that filed for patents were not very different. And much of the creative innovation occurred in the two least concentrated subsectors of food processing and rubber and plastics. Assuming that generating patents represents a higher level of innovative activity, relative to adoption or adaption (following the World Bank framework embraced in this study), this would suggest that technological capabilities were highest in the subsectors where many firms were competing with one another. This is contrary to the Shumpeterian idea that innovation increases with market concentration.

Gayle [17] offers a possible explanation; simple patent counts used to measure creative output fail to recognise the marked differences in technologies underlying these patents. Major innovations require significant resources that only larger firms can command, and they are more likely to be found in concentrated markets. In less concentrated markets, characterised by competition among many firms, product differentiation is common. And product differentiation can lead to numerous patents of minor changes to existing technologies or products; this will come across as a high level of creative activity among smaller firms in less concentrated industries when, in fact, they are merely patents for minor technologies arising from product differentiation.

Unfortunately, the survey data do not allow us to distinguish between patents underlying sophisticated technologies from patents due to product differentiation. But it is telling that food processing that had a large concentration of creators was among the activities described as being at "the low-tech end of the spectrum". In contrast, subsectors widely held to be utilizing more complex technologies such as electronics, machinery manufacturing and chemical products and which were also more concentrated, had smaller proportions of firms engaged in creative innovation. There is, therefore, some basis to conclude that much of the creative innovation observed among firms in the sample may be related to relatively unsophisticated technologies. If true, much more effort is needed to bolster innovations that are more sophisticated if manufacturing is to effectively pull the Malaysian economy up the value chain.

References

1. L. Einav, J. Levin, The data revolution and economic analysis. Innov. Policy Econ. **14**(1), 1–24 (2014)
2. D. Antons, C.F. Breidbach, Big data, big insights? Advancing service innovation and design with machine learning. J. Serv. Res. **21**(1), 17–39 (2018)
3. J.A. Doornik, D.F. Hendry, Statistical model selection with "Big Data". Cogent Econ Fin **3**(1), 1045216 (2015)
4. C. Kuhlman et al., How to foster innovation: a data-driven approach to measuring economic competitiveness. IBM J. Res. Develop **61**(6), 1–11 (2017)
5. S.H. Law, T. Sarmidi, L.T. Goh, Impact of innovation on economic growth: evidence from Malaysia. Malays. J. Econ. Stud **57**(1), 113–132 (2020)
6. S.M.P. Hosseini, S. Narayanan, Adoption, adaptive innovation, and creative innovation among SMEs in Malaysian manufacturing. Asian Econ. Pap. **13**(2), 32–58 (2014)
7. MASTIC, *Malaysian Science & Technology Indicators* (M.S.a.T.I.C. (MASTIC), ed.) (Malaysian Science and Technology Information Centre, Kuala Lumpur, Malaysia, 2008), p. 131
8. PICS, *Productivity and Investment Climate Survey 2, 2007, Enterprise Survey 2015* (W. Bank, ed.) (2015)
9. MPC (2011) *Productivity Report 2010/2011* (M.P. Corporation, ed.) (The Malaysian Productivity Council, Petaling Jaya), p. 300
10. H.W. Chesbrough, *Open Innovation: The New Imperative for Creating and Profiting from Technology*, ed. 1578518377, vol. 42 (Harvard Business Press, Boston, 2003), p. 231
11. D.A. Powers, Y. Xie, *Statistical Methods for Categorical Data Analysis*, 1 edn. (Emerald Group Publishing, United Kingdom, 2000), pp. 59–78
12. J.S. Long, J. Freese, Scalar measures of fit for regression models. Stata J. **1**(1), 1–9 (2000)
13. C. Fassio, F. Montobbio, A. Venturini, Skilled migration and innovation in European industries. Res. Policy **48**(3), 706–718 (2019)
14. OECD, *OECD Science, Technology and Innovation Outlook 2018* (2018)
15. C. Fassio, Exportled innovation: the role of export destinations. Ind. Corp. Change **27**(1), 149–171 (2017)
16. M. Woerter, S. Roper, Openness and innovation—home and export demand effects on manufacturing innovation: panel data evidence for Ireland and Switzerland. Res. Policy **39**(1), 155–164 (2010)
17. P.G. Gayle, *Market concentration and innovation: new empirical evidence on the Schumpeterian hypothesis*, in University of Colorado at Boulder: unpublished paper, (Kansas State University: Manhattan, Kansas, 2001), p. 35

Printed in the United States
By Bookmasters